Rainer Minixhofer

Integrating Technology Simulation into Semiconductor
Manufacturing

Rainer Minixhofer

Integrating Technology Simulation into Semiconductor Manufacturing

Bridging the Gap between TCAD and Semiconductor Manufacturing

VDM Verlag Dr. Müller

Imprint

Bibliographic information by the German National Library: The German National Library lists this publication at the German National Bibliography; detailed bibliographic information is available on the Internet at http://dnb.d-nb.de.

Cover image: www.purestockx.com

Publisher:
VDM Verlag Dr. Müller Aktiengesellschaft & Co. KG , Dudweiler Landstr. 125 a, 66123 Saarbrücken, Germany,
Phone +49 681 9100-698, Fax +49 681 9100-988,
Email: info@vdm-verlag.de

Zugl.: Wien, TU, Diss., 2006

Produced in USA and UK by:
Lightning Source Inc., La Vergne, Tennessee, USA
Lightning Source UK Ltd., Milton Keynes, UK
BookSurge LLC, 5341 Dorchester Road, Suite 16, North Charleston, SC 29418, USA

ISBN: 978-3-8364-7322-4

Sein und Wissen
ist ein uferloses Meer:
Je weiter wir vordringen,
um so unermeßlicher dehnt sich aus,
was noch vor uns liegt
jeder Triumph des Wissens
schließt hundert Bekenntnisse
des Nichtwissens in sich.

Isaac Newton

Abstract

INTEGRATED CIRCUIT MANUFACTURING is nowadays a multi-billion dollar business and one of the most complex industry branches in the world.

Historically the development costs for introducing a new technology generation (technology node) have been steadily increasing. This trend is even accelerated by entering the nanotechnology regime with its highly complicated new lithography processes. To keep this trend under control, Technology Computer Aided Design (TCAD) has gained more and more importance.

This "front loaded" approach in development has improved the speed and quality of the semiconductor process technology development significantly. It was also very successful in reducing the development cost significantly (by 35% in 2005[1]).

However TCAD is still mainly a very specialized tool for only a small group of engineers inside semiconductor companies. Despite the big savings potential, the consistent application of this methodology on the control of the manufacturing process is still in the fledgling stages.

Technology simulation has not done the leap, from a tool for a small group from specialists to a tool for a larger number of users in the semiconductor production area, yet.

Main problems of the broader use result from the complex, often little intuitive use of the tools, the complex underlying physical models and particularly by the missing integration into the work flow of a modern semiconductor production line.

This work aims to bridge the distance between the semiconductor manufacturing line and the TCAD group, which exists in nearly each semiconductor production company. Special attention has to be paid to the transfer of the operational work flow sequences to the TCAD system in a similar form.

This new system enables non-specialized engineers to profit from the advantages of a theoretical evaluation through the closely integrated TCAD framework.

ABSTRACT

In this work the overall situation of both worlds was analyzed and categorized in a structured, hierarchical way. A consistent and effective TCAD work flow was set up. The necessary information for this work flow was identified. Additionally the existing data interfaces between manufacturing and simulation were analyzed and their structure and coupling nodes were represented.

In the following the interfaces between TCAD and manufacturing, identified thereby, were subject to a more throughout analysis and finally an integrated interface system between the two "worlds" was implemented. A strong emphasis was put on the border condition that all available TCAD software packages could be used together with this new integrated interface system.

In order to be able to generate the necessary input information for simulations as automatically and resistant to errors as possible, based on this analysis a compact set of converters and data transfer procedures was defined. The interaction with the user of the TCAD system was limited to the absolutely necessary minimum. This led to a strong improvement of the quality, reliability and also predictability of the results of the TCAD simulations.

The converters were integrated into the entire TCAD work flow. Several examples of nearly each aspect of the typical TCAD work flow show the most important effects of this new approach. In addition, during the development of new process technologies, particularly for the support of typical production problems (like yield problems in the manufacturing) this approach has been tested successfully. Furthermore, optimizations of furnace programs (a task which may frequently occur during manufacturing of qualified processes too) have been performed. The use of diffraction correction calculation for a better representation of interconnect structures could be shown. Thus a substantial improvement in the computation of parasitic elements, the optimization of an EEPROM cell and the three-dimensional simulation of a lateral pin diode could be obtained. As a further application area the inverse modeling of a polycrystalline fuse was shown. The difficult to measure thermoelectric material parameters of the materials used in the fuse (e.g. tungsten, titanium, titanium nitride) were determined through inverse modeling. Finally the use of technology simulation within the area of statistic process control was demonstrated.

The work closes with a short outlook and open problems.

Acknowledgment

F IRST AND FOREMOST I want to thank Prof. SIEGFRIED SELBERHERR for giving me the opportunity to conduct my scientific work in parallel to my professional engagement in semiconductor industry. Furthermore, for providing the excellent infrastructure at the Institute for Microelectronics, and for the professional support and help in getting me to finalize my thesis.

I thank Prof. WOLFGANG PRIBYL, that he was willing to serve on my examination committee even on very short notice.

I am grateful to my boss MARTIN SCHREMS to help me with my thesis in a very supportive and positive manner. Furthermore, he provided a very innovative and interesting working environment.

I wish to thank all my co-authors for the very good and fruitful discussion on the contents of my publications.

I especially want to thank GEORG RÖHRER from austriamicrosystems AG, DIETER RATHEI from DR Yield, STEFAN HOLZER and ANDREAS HÖSSINGER from Institute for Microelectronics, TU Vienna and JÜRGEN LORENZ from the Fraunhofer Gesellschaft, IISB for their contributions to the examples in Chapter 6.

After all my wife CHRISTINE and my children CHRISTOPH and BENJAMIN provided me with love and understanding, especially when another weekend passes by, me sitting at the computer and working on my thesis. They carried me through the tough times last year, when the work seemed to be never ending.

Finally, none of my studies would have been possible without the continuous support of my parents.

This work is dedicated to my father
WERNER MINIXHOFER
who passed away far too early

'... *Das Unmögliche = Das, was man nie versucht hat.*'
(The impossible = What one has never tried)

Hans Günther Adler (1910 - 1988)

Contents

List of Abbreviations and Acronyms

AAPSM	...	Alternating Aperture Phase Shift Masks
ASCII	...	American Standard Code for Information Interchange
NMOS	...	n-type **MOS**
PMOS	...	p-type **MOS**
BiCMOS	...	Bipolar & Complementary **MOS**
BSIM	...	Berkeley Short-Channel IGFET Model
CD	...	Critical Dimension
CIF	...	Caltech Intermediate Format
CMOS	...	Complementary **MOS**
CMP	...	Chemical Mechanical Polishing
COG	...	Chrome on Glass
DFT	...	Design for Test
DOE	...	Design of Experiments
dpm	...	Defective Parts Per Million
DRAM	...	Dynamical **RAM**
DUT	...	Device Under Test
EAPSM	...	Embedded Attenuated Phase Shift Masks
ECAD	...	Electronic Computer-Aided Design
EDA	...	Electronic Design Automation
EEPROM	...	Electrically Erasable Programmable Read-Only Memory
FET	...	Field-Effect Transistor
GDSII	...	Geometric Data Stream II
HTM	...	Half Tone Masks
HTML	...	Hypertext Markup Language

IC	...	Integrated Circuit
I/O	...	Input/Output
ITRS	...	International Technology Roadmap for Semiconductors
IV	...	Current(I)-Voltage
LDD	...	Lightly Doped Drain
LOCOS	...	Local Oxidation of Silicon
LSL	...	Lower Specification Limit
MES	...	Manufacturing Execution System
MOS	...	Metal-Oxide-Semiconductor
MPU	...	Microprocessor Unit
MPW	...	Multi Product Wafer
MOSFET	...	MOS Field-Effect Transistor
NTRS	...	National Technology Roadmap for Semiconductors
NVM	...	Non-Volatile Memory
OPC	...	Optical Proximity Correction
OTP	...	One Time Programmable Device
PCM	...	Process Control Monitor
PCI	...	Process Capability Index
PERL	...	Practical Extraction and Reporting Language
POR	...	Process of Record
ppb	...	Parts Per Billion
ppm	...	Parts Per Million
RAM	...	Random-Access Memory
RIE	...	Reactive Ion Etching
RTA	...	Rapid Thermal Annealing
SDT	...	Single Die Tooling
SIMS	...	Secondary Ion Mass Spectroscopy
SLM	...	Scribe Line Monitor
SMU	...	Source Measure Unit
SOC	...	System On a Chip
SPICE	...	Simulation Program with Integrated Circuit Emphasis
SPR	...	Simple Process Representation
SRAM	...	Static Random-Access Memory
TCAD	...	Technology Computer-Aided Design
TEM	...	Transmission Electron Microscopy
USL	...	Upper Specification Limit
UV	...	Ultraviolet
VBIC	...	Vertical Bipolar Intercompany Model
VHDL	...	Very High Speed Integrated Circuit Hardware Description Language
WIP	...	Work in Process

List of Symbols

Notation

x	...	Scalar
$\vec{x}$	...	Vector
$\vec{e}_x$	...	Unity vector in direction x
$\vec{e}_n$	...	Unity vector in direction of vector $\vec{n}$
$\vec{x} \cdot \vec{y}$	...	Scalar (in) product
$\vec{x} \times \vec{y}$	...	Cross product
$\frac{\partial(\cdot)}{\partial t}$	...	Partial derivative with respect to t
∇	...	Nabla operator
$\nabla \vec{x}$	...	Gradient of $\vec{x}$
$\nabla \cdot \vec{x}$	...	Divergence of $\vec{x}$
$\nabla \cdot \nabla = \nabla^2$	...	LAPLACE operator
$\langle \cdot \rangle$	...	Statistical average
$f(\vec{r}, \vec{k}, t)$	...	Distribution function
$\mathcal{F}, \mathcal{F}^{-1}$	...	FOURIER transform and inverse FOURIER transform respectively
$J_n(\cdot)$	...	BESSEL function of nth order

Constants

h	...	PLANCK's constant	$6.6260755 \times 10^{-34}$ Js
$\hbar$	...	Reduced PLANCK's constant	$h/(2\pi)$
k_B	...	BOLTZMANN's constant	1.380662×10^{-23} JK^{-1}
q	...	Elementary charge	$1.6021892 \times 10^{-19}$ C
m_0	...	Electron rest mass	$9.1093897 \times 10^{-31}$ kg
κ_0	...	Dielectric constant	$8.8541878 \times 10^{-12}$ AsV^{-1}m^{-1}
$\imath$	...	$\sqrt{-1}$	

List of Figures

List of Tables

'... The complexity for minimum component costs has increased at a rate of roughly a factor of two per year. Certainly over the short term this rate can be expected to continue, if not to increase. Over the longer term, the rate of increase is a bit more uncertain, although there is no reason to believe it will not remain nearly constant for at least 10 years. That means by 1975, the number of components per integrated circuit for minimum cost will be 65,000. I believe that such a large circuit can be built on a single wafer.'

Gordon E. Moore, 1965

Chapter 1

Introduction

Semiconductor technology and industry has enormously advanced in the past decades. Starting from a plastic triangle, a slab of germanium, some gold foil and gold contacts (the first bipolar transistor in 1947), as of 2004 the typical transistor density per circuit is around 140 million transistors/cm^2 for MPU (micro processor unit) applications, doubling every year (Moore's Law [2]). This trend is shown in Figure 1.1 starting with the 4040, the first Intel Processor in 1971. Semiconductor Industry is the main driving force for technology innovation and "New Economy" markets. The ongoing development of faster integrated circuits with higher device density has led to highly complex and sophisticated products which are widely accepted by society. A modern integrated circuit cannot be developed without the massive use of computer aided design (CAD) in any step of the complex flow from the idea to the final product. This work concentrates on technology computer aided design (TCAD) [5] and its integration into the semiconductor fabrication process flow. The use of TCAD is twofold: Firstly it models the complex flow of semiconductor fabrication steps ending up with detailed information on geometric shape and doping profile distribution of a semiconductor device in scope (like CMOS- or Bipolar-Transistors) ⇒ Process TCAD-Simulation. Secondly it uses the information of the first step to predict the device characteristics of semiconductor devices leading to circuit simulation models as implemented in any circuit simulator like PSPICE [6], ELDO [7], SPECTRE [8] etc. ⇒ Device TCAD-Simulation. The setup of such a simulation methodology requires an almost completely documented semiconductor fabrication process flow including such fabrication details like angle of incidence of ions implanted in ion implantation process steps, or etch rate distribution as a function of the local angle of the etched layer surface. Any modern semiconductor fabrication facility maintains such documentation to an extremely high detail level, but commercial TCAD simulation software like Synopsys [9] or Silvaco [10] Tools need this information in a very specialized format [11] which cannot be directly deduced from the standard process flow documentation. The traditional way of setting up the process- and to some extent also the device TCAD-simulation framework is, entering it by hand, which is of course a source of numerous errors. This work proposes a new methodology with the main target to automate this conversion process to a high extent.

Chapter 2 describes the overall chain of processes, how integrated circuits are fabricated. An overview with respect to the related simulator tools is given.

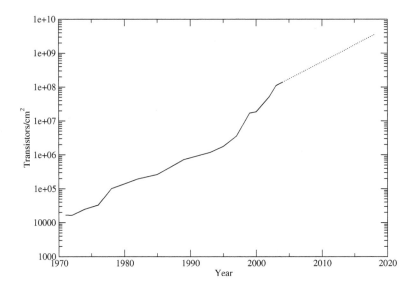

Figure 1.1: Intel MPU transistor density trend starting with the 4040 processor [3]. The dashed line shows the ITRS chip size model [4]

Chapter 3 concentrates on the detailed simulation methodologies to model the IC (Integrated Circuit) fabrication.

Chapter 4 identifies the interfaces between the fabrication process and simulation. It outlines the detailed structure of the interfaces. A comprehensive overview over the interactions in this integrated system is given as well.

Chapter 5 provides the detailed description of how this interfaces are implemented.

Chapter 6 demonstrates the strengths of such a structured and integrated approach with a couple of cases in a real semiconductor fabrication environment.

Finally, **Chapter 7** briefly summarizes this work with some conclusions and an outlook.

'... The bourgeoisie has stripped of its halo every occupation hitherto honoured and looked up to with reverent awe. It has converted the physician, the lawyer, the priest, the poet, the man of science, into its paid wage labourers.'

Karl Marx and Friedrich Engels, 1888

Chapter 2

The Processing Chain in Semiconductor Manufacturing

2.1 Overview

The semiconductor industry is starting from the product idea the following sequential steps occur in a standard integrated circuit development and production flow [12].

1. Development: Starting from the product idea, the electronic contents of the overall system are developed, leading into a schematic of the electronic circuitry. For digital circuitry this development process is similar to writing a software program by using *Very High Speed Integrated Hardware Description Language* (VHDL) as a abstract description of the digital block. The development of a digital block starts with the specification (operation and timing) and the subsequent description of this specification via a model in VHDL.

2. Design: The integrated circuit is designed starting from the schematic, and taking into account the special demands of integrated circuits (crosstalk, common substrate, etc.). It is now standard to use ECAD tools to simulate the behaviour of a design as an integrated circuit by using detailed circuit simulation models and design rules, which are specific to a process family (technology node) [13],[14].

3. Layout: The resulting integrated circuit is drawn as a layout on the specific layers which are given by the semiconductor process family (technology node). The combination of multiple layers, like implantation masks and etch masks, define the shape and functionality of the electronic devices in the integrated circuit [15],[16],[17].

4. Mask-Shop: The layout is post processed to take into account process induced size variations (layer biasing) and constraints on combination of layers (logical combination). The physical mask layers are written from this data by using laser- [18] or e-beam [19] equipment.

5. The wafer start material is released at the beginning of the process flow into fabrication [20]. In the following these wafers are subject to numerous single process steps like

ion-implantation, deposition and etching of semiconductor, dielectric, and metallic materials, furthermore diffusion of dopants, and oxidation and lithography to structure deposited layers using the previously fabricated mask reticles.

6. After leaving the fabrication the now functional integrated circuits are tested electrically. Firstly on single device level on process control monitors (PCM's), secondly on full device level (wafer sort). These tests select the functioning parts for further processing.

7. Scribing into pieces and packaging of the single circuits.

8. Electrical functionality test of the packaged pieces.

The overall processing chain is shown in Figure 2.1.

The ECAD simulation tools in Subject 1, Subject 2, and Subject 3 are already closely integrated into the development chain [21] and are therefore very efficient.

Packaging simulation is not subject to this work, however, tools [22], [23] are used to analyse new packages with respect to electromagnetic field, stress, and self heating.

For Subject 4 to Subject 6 good simulation solutions exist for the single process step (e.g. SIGMA-C or PROLITH for lithography and mask fabrication step simulation, TCAD Tools from Synopsys and Silvaco for the process- and device-simulation steps), which are sufficient for most of the two-dimensional process- and device-simulation applications.

However, the set-up of these TCAD-simulators is highly complicated and time consuming. Changes in fabrication procedures like parameter optimization of process conditions are not reflected in simulation with the traditional way of defining this set-up by hand. Therefore the simulation flow definitions become asynchronous to the semiconductor fabrication very quickly.

The main concept to be considered is to match the simulation methodology as closely as possible to the fabrication methodology in an automatic (or at least semi-automatic) way. The resulting work flow and the main application areas for TCAD integration into fabrication can be seen in Figure 2.2.

In the following the main aspects of the parts of this implementation are outlined.

2.2 Design

As mentioned above the integration between design of integrated circuits and their simulation is already very efficient. The circuit design is almost exclusively performed at workstations by using a sophisticated set of software tools to model the behaviour of the schematic to be implemented as integrated circuit on silicon wafers. To enable a high degree of modeling accuracy an extensive set of characterization of the available devices on silicon has to be done in advance. This task is called *process and device characterization (PDC)*. It will be shown in Chapter 4 how TCAD may support this task especially for the development of new process technologies. PDC is generating the model parameters for the SPICE models used by circuit designers. The consistency and accuracy of these SPICE models is absolutely mandatory for enabling designs with the envisaged electrical specifications. The generation of the SPICE models is carried out after the

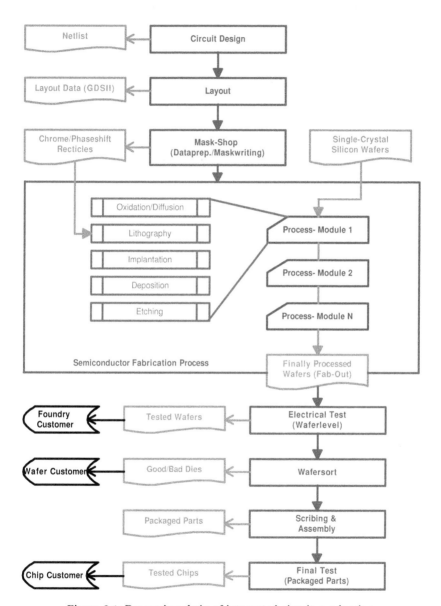

Figure 2.1: Processing chain of integrated circuit production

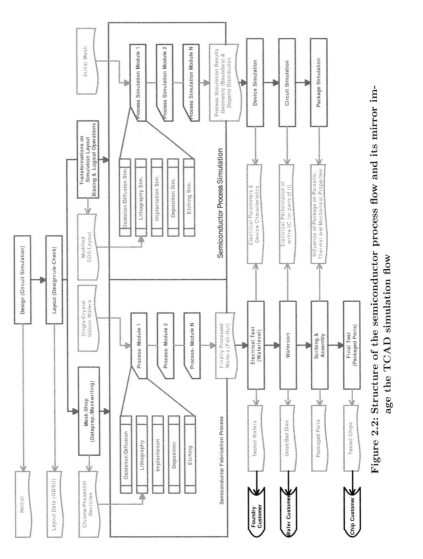

Figure 2.2: Structure of the semiconductor process flow and its mirror image the TCAD simulation flow

semiconductor process freeze during development of new process flows or occasionally, if some major semiconductor process change occurred. To reflect shifts or drifts in the semiconductor device performance as well as the statistical variations of the process technology a feedback loop is established between electrical test and the SPICE models. This feedback is implemented by using the pass/fail limits of the electrical test to supply worst/case conditions to the designer. There are numerous new approaches for simulating statistical fabrication fluctuations within the ECAD environment labeled under the term DFM (design for manufacturability) [24],[25], however, the details are outside of the scope of this work. By using SPICE models, EDA (Electronic Design Automation) design tools from companies like Mentor [26], Cadence [27], or Agilent [28] are able to simulate the behaviour of the schematic entered by hand or imported with net lists. A typical design flow is shown in Figure 2.3.

2.3 Layout

After finalization of the behavioural modeling of the schematic, the design is layed out into a complex combination of masks. There are several basic building blocks of the layout. The digital library consists of predefined highly optimized (in terms of speed and area) digital cells which are connected automatically by metal interconnect layers during place and route.

Memory blocks (SRAM, DRAM, OTP or NVM) are normally placed by memory generators automatically. The most demanding building block for a designer is the analog part of a design, where linearity, accuracy, and matching of the single devices play an important role. Especially the layout of this block may have an important influence on the subsequent performance of the analog part of the design by introducing additional parasitics between the single devices in the silicon (e.g. leakage). High frequency or timing dependent applications can be strongly influenced by the RLC-network of the interconnects between the analog components. Therefore, after layout a parasitic extraction has to be performed, where the RLC-network of the interconnects is extracted from the layout and this extracted values are put into the overall schematic of the integrated circuit in an additional circuit simulation run. Finally the bonding pads are placed to provide the ports of the design, where the package bond wires are connected to. There are several different pad types:

1. Digital I/O pads: These pads are optimized for the connection to and from the digital logic. The operating voltages range from 5V down to 1.3V for low power logic.

2. Analog I/O pads: These pads take special measures to provide different voltage levels depending on the process technology [29] (e.g. for high voltage drivers analog outputs have to support higher voltages than 5V).

3. VDD and VSS pads: These pads are for the supply of the chip.

The main difference of the above types is in their ESD (electrostatic discharge) protection concept. Because the devices connected to the pad types are very different (digital cells, drivers or power buses), the ESD protection concept has to account for this independently. A special subset of the Analog I/O pads is frequently termed as RF pad which provides a reduced amount of interconnect parasitics for critical analog high frequency applications. After finalization of the layout the integrated circuit may be *clustered* together with other circuits, if multiple products

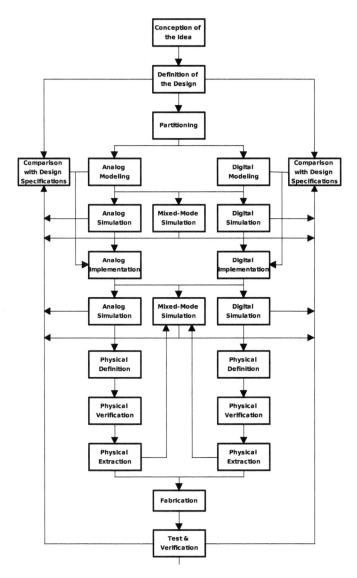

Figure 2.3: Design flow of a mixed-signal design, comprising of an analog and a digital part

are processed together in one batch in an MPW (multi product wafer batch). For high volume products a SDT (single die tooling) is prefered, where only one product is processed in a wafer batch. The *tape out* marks the transfer of the layout data to the mask shop for generation of the lithography masks for processing. The data can be transferred in different formats like CIF, GDSII, EBES, or EDIF which are described in more detail in Appendix B.

2.4 Mask Generation

2.4.1 Introduction

A fundamental requirement for almost all useful semiconductor devices is the definition of patterned elements. The main stream technology choice for patterning has been optical lithography. Up to the early 70's lithography was done as a contact printing process in which blue or near UV light was passed through a photo mask directly onto a photo resist coated semiconductor substrate [30].This shadow imaging process has been described in many research publications and handbooks [31], [32]. Beginning in the early 80's a new class of projection exposure tools, known as steppers, was introduced [33]. For the first time the pattern definition imaging on the semiconductor wafers was performed one chip at a time in a step-and-repeat fashion. Most stepper systems employed a reduction projection lens to ease the fabrication difficulty of the photo mask and to improve the overall precision and accuracy of the overlay of patterns on the wafer. Even more recently a combination of the earlier scanning approach with the step-and-repeat approach was created [34]. The step-and-scan approach has spread rapidly throughout the lithography tool industry, and is used for critical layers (like gate, metallization and contact layers) at the 250nm node and below [35]. Until the mid 90's all optical photo masks have been chrome on glass (called COG-photo masks) [36], also called binary photo masks. Starting from the 350nm node significant innovations in binary masks such as OPC (optical proximity correction) [37] and AAPSM (alternating aperture phase shift masks) [38] were introduced which improved the resolution capability of binary photo masks. A second approach besides the binary photo masks then emerged as EAPSM (embedded attenuated phase shift masks) [39] also called HTM (half tone masks).

2.4.2 Imaging Basics

Lithography is based on replicating the pattern on a photo mask into resist covered wafers. In an ideal case without degradation in the imaging process, a simple copy of the mask pattern would result, as shown in Figure 2.4 a.

However, in a projection process the imaging is always subject to degradation from diffraction and from imperfections in the projection system. An example of the image from a diffraction-limited projection system is shown in Figure 2.4 b. The spreading of the image profile results from the wave nature of light, and it is this property that limits the resolution capability of optical imaging systems. In an imaging lens system with a circular aperture of radius $\tilde{r}_0$ and imaging distance D', the image intensity resulting from a point source can be described by an expression containing a first order BESSEL function,

$$I'(x) = I(0) \left(2 \frac{J_1(x)}{x} \right)^2 \tag{2.1}$$

where $x = \frac{2\pi \tilde{r_0} r'}{D' \lambda}$ and r' is the distance in the image plane from the geometrical image point. λ is the wavelength of the monochromatic light source. A detailed deduction of this expression is given in Appendix E.2. The fraction $\frac{\tilde{r_0}}{D'}$ given by

$$\frac{\tilde{r_0}}{D'} = \tan \theta \cong \sin \theta \qquad \text{for} \qquad \theta \ll \tag{2.2}$$

equals the numerical aperture NA defined by

$$NA \equiv n \sin \theta \tag{2.3}$$

with n as the refractive index of the medium behind the aperture or lens. Therefore the expression for x can be further simplified to $x = 2\pi \rho \frac{NA}{\lambda}$. For air as medium ($n \approx 1$) a simplified description of NA is given in Figure 2.5 as $NA = \sin \theta$

This light intensity distribution is known as the AIRY pattern, after G.B. AIRY who first derived it in 1835 [40]. In addition to the general shape of the curve, shown in Figure 2.6, the first zero value is of interest. At about $x = 0.612\pi$ occurs a intensity minimum and an intensity maximum at $x = 0.822\pi$.

Resolution is defined as the ability to distinguish components of an object or a group of objects. The resolution capability of astronomical telescopes was studied in detail by LORD RAYLEIGH in the 19th century [41]. He defined the limit of resolution for a telescope as the angular separation between two stars when the peak of the AIRY intensity pattern from one star coincided with the first minimum of the AIRY intensity pattern for the other star. This leads to the well-known RAYLEIGH condition for angular resolution

$$NA = \sin \theta = 0.61 \frac{\lambda}{\tilde{r_0}} \tag{2.4}$$

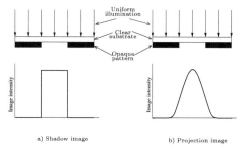

a) Shadow image   b) Projection image

Figure 2.4: Basic imaging characteristics (a) Ideal shadow imaging; (b) Diffraction-broadened projection imaging

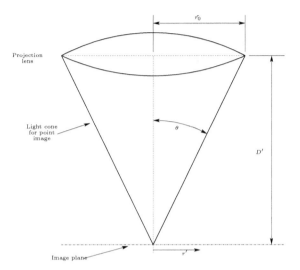

Figure 2.5: Geometric situation in simple projection optical system

where $\tilde{r}_0$ is again the radius of the imaging objective aperture. A sketch of the RAYLEIGH resolution condition is shown in Figure 2.7. Note that the intensity at the midpoint between the image peaks is reduced to about 78% of the peak intensity, which provides discernible separation, but not with high contrast between the bright and dark regions.

While the analogy of astronomical imaging to photo lithography is not completely quantitative, some key observations can be made. There is a limit to resolution for any given optical projection system, and it is not possible to resolve arbitrarily small or closely spaced features. It is also apparent that the resolution can be improved by using a smaller wavelength of the exposure light, and the resolution can be improved by making the projection system aperture larger.

In practical lithography the RAYLEIGH condition is typically restructured into the "RAYLEIGH equation"

$$resolution = k_1 \frac{\lambda}{NA} \tag{2.5}$$

where NA is the numerical aperture of the projection system and k_1 is a constant in the order of 0.4-0.8. There is no rigorous optical definition for the constant k_1, and it is generally used as a qualitative descriptor of the overall lithography process capability. More details on the RAYLEIGH equation are given in Table 2.1. Table 2.2 shows the numeric aperture, the resolution, and the depth of focus for the most important wavelengths in lithography.

This common description of resolution capability is closely related to the AIRY pattern described above. In particular, the first minimum of the AIRY pattern occurs at about $r' = 0.61 \frac{\lambda}{NA}$, and

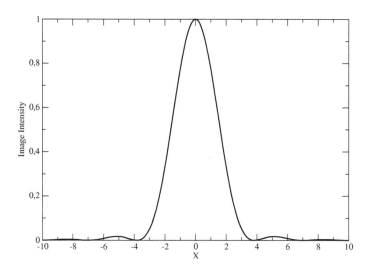

Figure 2.6: Light intensity distribution from a point source projected through a circular imaging lens. The variable x on the horizontal axis is defined in the text

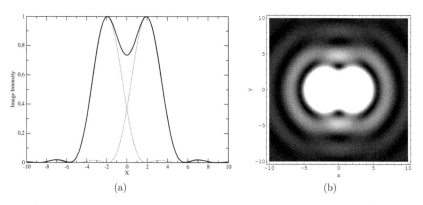

(a)   (b)

Figure 2.7: RAYLEIGH criterion for resolution of two point images in (a) 1D and (b) 2D. Scale is the same as in Figure 2.6

Definitions

λ(g-line)	436 nm
λ(i-line)	365 nm
λ(KrF)	248 nm
λ(ArF)	193 nm
$\lambda(F_2)$	157 nm

Resolution

RAYLEIGH Resolution	$R = k_1 \frac{\lambda}{NA}$
Traditional	$k_1 = 0.8$
Advanced	$k_1 = 0.4 - 0.6$

Depth of Focus

RAYLEIGH Depth of Focus	$DoF = k_2 \frac{\lambda}{NA^2}$
Traditional	$k_2 = 1.0$

Table 2.1: **Definitions for important wavelength nodes in lithography**

Wavelength	NA	Resolution[μm]	DoF [μm]
i-line	0.63	0.35	0.92
KrF	0.60	0.25	0.69
KrF	0.70	0.21	0.51
ArF	0.70	0.17	0.39
F_2	0.70	0.13	0.32

Table 2.2: **Examples for typical lens configurations ($k_1 = 0.6$) for deep sub micron technology nodes**

the first maximum occurs at $r' = 0.82 \frac{\lambda}{NA}$. The qualitative agreement with the usual range of k_1 is apparent.

2.4.3 Optical Proximity Effect

High performance optical projection imaging for lithography is strongly impacted by diffraction effects as noted in several previous sections. One result of this behavior is that individual pattern features do not image independently, but rather they interact with neighboring pattern features. A detailed analysis of the projection imaging process, for example, the analysis described in the paper by Hopkins [42], considers contributions from every portion of the reticle object and every portion of the projection optics in determining the exact image at the wafer plane. A simple heuristic argument considers the extended diffraction structure of the AIRY function comprising of the additional local maxima in the intensity distribution. Overlap of the diffraction peaks with adjacent pattern features leads to increased or decreased exposure intensity at any point in the image, compared to a purely geometrical image model.

2.5   Fabrication

2.5.1   Description of Semiconductor Manufacturing Processes

In the following subsections an overview over the different process steps, a wafer undergoes during its fabrication in the clean-room, is given. A semiconductor manufacturing process differs markedly from other processes. In many other types of processing plants, the material being processed moves through the plant in a fairly simple, straightforward, and well-integrated manner. Despite the fact that the processing flow of this material is straightforward and linear, a flow chart depicting the process will usually be quite complicated. Contrast this with a semiconductor manufacturing process, which can be described very easily with a linear processing flow chart, but whose work-in-process (WIP) moving through the plant will follow complex paths, crisscrossing back and forth in intricate patterns. During wafer processing - i.e. in the semiconductor fabrication clean-room - the integrated circuitry is formed at the surface of the single crystal silicon wafer by numerous repetitive micro-lithographic, deposition, diffusion, and etching steps, until it is finished. During this processing, depending on the complexity of the technology, a set of about fifteen up to more than thirty-five separate wafer processing cycles (which form modules like gate module, LDD module, metal module and so on), including the associated lithography step, were performed. An expanded flow chart of one of these cycles appears as shown in Figure 2.8.

Here it can be seen that the wafer will iterate through this inner circle as many times as there are masks(alignments)[1] for adding new circuitry.

2.5.2   Lithography

Micro-lithography is the process of defining useful shapes on the surface of a semiconductor wafer. Typically this consists of a patterned exposure into some sort of photosensitive material already deposited on the wafer. A variety of processes that directly pattern the wafer are possible, such as directly writing on the wafer with an electron beam, or nanoimprinting structures with stamps, but at this time none is in use for high volume semiconductor manufacturing.

The imaging basics of optical lithography have been outlined in Section 2.4 already. Figure 2.9 shows the principal components of the illumination system of a typical i-line stepper [43] practical for structure sizes down to 350nm.

A detailed description of the lithography process and its modeling basics can be found in [44].

2.5.3   Ion Implantation

Ion implantation is a process whereby energetic dopant ions are made to impinge on a silicon or other target, resulting in the penetration of these ions below the target surface and thereby giving rise to controlled, predictable dopant distributions. Low implant energy produces dopant distributions near the surface such as are required for MOS source and drain regions, or bipolar

---

[1]One has to carefully distinguish between the terms "mask" and "alignment". A mask is the physical reticle for the illumination process. An alignment is the group of steps performed in a lithography track as shown in Figure 2.8 in the box. The number of masks and alignments is normally not equal, because certain reticles may be used for more than one alignment.

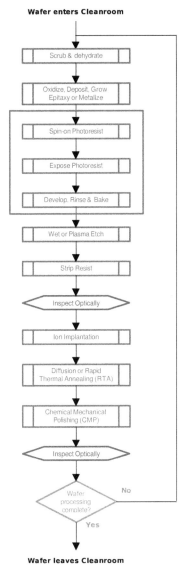

Figure 2.8: Expanded flow chart of the wafer fabrication cycle comprising one alignment step and the associated processing

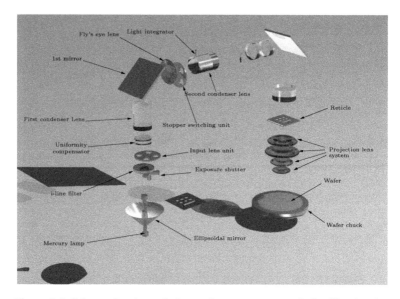

Figure 2.9: Schematic view of the main components of the illumination system of a lithography tool

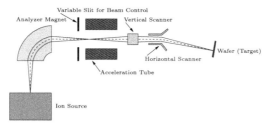

Figure 2.10: General schematic of an ion implantation equipment

emitter regions. High energies produce deeply implanted dopant profiles as required for CMOS retrograde wells and buried layers. The provision of wafer cooling during implant allows the use of photo resist masks to laterally control the location of dopant regions, an inevitable feature for the production of CMOS devices. Similarly, topographical features of the device, such as a gate stack, may be used to impose additional masked regions to the implant, thereby allowing for the production of cost- and yield-effective, self-aligned doping regions. A detailed description of the physics of this process step can be found in, e.g., [45],[46],[47],[48]. The general implanter schematic is shown in Figure 2.10.

2.5.4   Diffusion

Diffusion is a key task of semiconductor wafer processing. Although dopants are generally introduced into a wafer by ion implantation, rather than thermally in a furnace, there is unavoidable diffusion of the dopants during any high temperature process step. The models in this area can be categorized into two major approaches, namely, the continuum theory of Fick's diffusion equation and the atomistic theory. The continuum theory requires the solution of Fick's diffusion equation, generally with constant values for the diffusion coefficient and is adequate for low dopant concentrations. When the doping concentrations are high, the diffusion profiles may exhibit anomalous diffusion behavior and a simple form of Fick's law cannot be applied, because the diffusion coefficient becomes concentration dependent. The picture then becomes considerably more complicated and requires an atomistic approach which studies the interactions between native point defects (vacancies and interstitials) and dopant atoms. The underlying idea behind all this is that the dopant atoms mostly dissolve substitutionally in the lattice. Only through interactions with native point defects are the dopant atoms able to jump form one site to another, effecting long-range diffusion.

Due to the agitation of the lattice by phonons some of the defects can wander throughout the lattice. For a simple cubic lattice this diffusion of defects can be understood by considering the jump process between two adjacent (100) planes, 1 and 2. If the lattice planes contain n_1 and n_2 defects per unit surface area, respectively, and the jump rate in either direction is given by ν then the number of defects per unit surface area jumping form plane 1 to 2 in time dt is $J_1 dt = n_1 \nu dt$. For the same jump probability in either direction, the net flux of dopants from plane 1 to 2 J_{12}, can be written as

$$J_{12} = J_1 - J_2 = (n_1 - n_2)\nu \tag{2.6}$$

For a small lattice constant a and assuming that the number of defects changes slowly with distance x (continuum approximation), the above flux can be written as

$$J = -\nu a \frac{\partial n}{\partial x} \tag{2.7}$$

The defect concentration per unit volume $C = (n/a)$, (2.7) becomes

$$J = -a^2 \nu \frac{\partial C}{\partial x} = -D \frac{\partial C}{\partial x}, \tag{2.8}$$

where $D = a^2 \nu$ is the diffusion coefficient or diffusivity. The above equation is Fick's first law of diffusion.

$$D = D_0 \exp -\frac{Q}{k_B T} \tag{2.9}$$

where Q is the activation energy. Q is just H_m^X, and D_0 is proportional to the Debye frequency. A more detailed description of the theory of diffusion in semiconductors can be found in [49],[50] and [51].

2.5.5   Film Deposition

A variety of materials can be deposited by chemical vapor deposition (CVD). Some typical materials are silicon nitride (Si_3N_4), silicon dioxide (SiO_2), TEOS ($Si(OC_2H_5)_4$), polycrystalline silicon, and various metals. Depending on the particular deposition method, the temperature varies from about $300\,°C$ up to $900\,°C$. Additionally the pressure range may vary significantly which differentiates atmospheric pressure chemical vapor deposition (APCVD) from low pressure chemical vapor deposition (LPCVD). A typical characteristic of the deposition process is the deposition rate [nm/min]. Film deposition is used to deposit other materials on top of the silicon wafer. These materials are necessary to build functional parts of the devices (PMOS, NMOS,Bipolar Transistor) and their interconnects in the integrated circuit.

For simulation, deposition is performed by geometry operations, where the deposition rate may vary locally. Details about different simulation approaches for deposition can be found in [52], [53] and [54].

2.5.6 Etching

Etching can be subdivided into two main categories, isotropic and anisotropic etching.

2.5.6.1 Isotropic Etching

This category describes etching rates which are independent of direction. Isotropic etching is usually performed by means of wet chemistry and the wafer is immersed into a reactive solution. The etchant species diffuses towards the wafer surface, dissolution of the specific material takes place, and the generated products separate in turn by diffusion from the surface [55]. The advantages of wet etching are the possible high selectivity for specific materials and the low damage to the substrate.

2.5.6.2 Anisotropic Etching

A typical anisotropic method is reactive ion etching (RIE) [56]. Here ions are accelerated through a Chlorine- or Fluorine-based-plasma towards the wafer surface. The Chlorine (or Fluorine) penetration into the silicon surface is strongly enhanced by the ion bombardment. Since vertical surfaces are less exposed to the ions, a large ratio in the etching rates can be obtained. Furthermore (especially for deep trench etching), passivation of the etched sidewalls is obtained by polymerization of some components of the etching chemistry at the sidewalls. Thus only the bottom of the advancing etching front which is exposed to the ion bombardment is free of these polymers and, therefore, a strong anisotropy of the etched structure can be obtained. One drawbacks of the method is the possible damage to the silicon substrate due to the high energy of the argon ions. Furthermore, the excitation of the plasma field with an RF electromagnetic field induces potentials in the interconnect wires contacting e.g CMOS transistor gates and subsequently damaging the gate oxide of the CMOS transistors (antenna effect).

2.6 Electrical Test

To evaluate the quality and stability of the semiconductor manufacturing process a couple of electrical parameters are measured, at the stage, when the wafers are finished with processing (fab-out). Every semiconductor factory uses PCMs (Process Control Monitors) for this purpose.

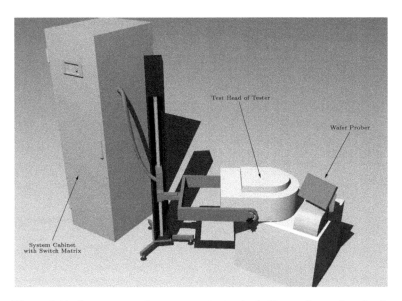

Figure 2.11: An automated parameter tester including wafer prober for final wafer acceptance test

A PCM is normally positioned in the scribe line between the integrated circuits [57]. The scribe line is the area where the identical dies are sawed [58] before the integrated circuits are assembled in packages. These scribe lines have a width between 150 and 60 microns and a typical length of the lithography step field (approx 2 mm). Since the available width is very small, complicated circuitry cannot be used inside a PCM. Normally the test structures are comprised of single transistors, resistors, capacitors, and other passive structures.

The commonly used measurement equipment for this task is a parametric tester including an automated wafer prober for fast and automated handling.

To be able to measure the small control structures the wafer prober must be able to use pattern recognition to align exactly on the probe pads of the PCM structures. Furthermore, it has to move to the structures in a fast an reliable way. Typical commercial equipment for such a prober offer a standardized interface for control of the wafer handling via the electrical parameter tester. The parameter tester consists of a group of high-accuracy and fast measurement equipment and power source units so called "SMU"'s (Source Measure Units) which are able to measure voltage differences down to the μV range and currents in the range of picoamps [59]. Furthermore, they offer integrated voltage and current sources which can be programmed in a very flexible way. These SMUs can be connected to a solid-state switch matrix which wires these instruments to a probe-card mounted in the wafer handler that connects to the probe-pads of the PCM structures during electrical test. An overall schematic of such an automated system is shown in Figure 2.11.

The typical wafer test steps are as follows (see Figure 2.12):

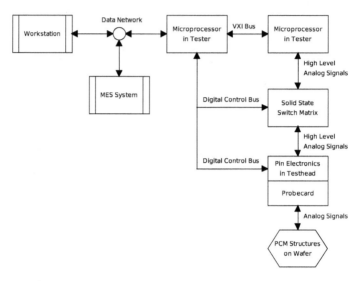

Figure 2.12: Schematic overview of how the automated parameter tester system performs a test

1. The control computer (mainly a UNIX workstation) sends a test program to the controller in the system cabinet through a data network connection.

2. The controller converts the test information for use of the system.

3. Test data is sent to the SMU in the system cabinet.

4. The SMU sends the test measurement requirements to the probe in the test head.

5. Test results are measured by the SMU or a capacitance meter.

6. The results are sent through the data network connection to the control computer for processing.

The PCM structures are measured in terms of electrical characteristics and certain parameters are extracted for monitoring purposes. The parameters form a hierarchy of parameter classes, depending on the importance of their value distribution for integrated circuits. There are three classes of parameters shown in Table 2.3.

For the class of pass/fail parameters there are defined specification limits. These limits reflect the specifications a process technology has to fulfil to enable competitive integrated circuit designs, e.g., if one has to cope with a very high variation of threshold voltage of MOS transistors one has to use big transistors to achieve a certain stability of his design. This area consumption limits the competitiveness of the product and therefore of the entire process technology. This constraint leads to tight parameter limits which have to be controlled in an active way to ensure

Parameter Class	Purpose	Measurement Frequency
Pass Fail	Defines if wafer material is acceptable	At least 5 PCM monitors on every wafer
Information	Provides additional statistical information on device behaviour	At least 5 PCM monitors on every wafer
Charact- erization	Gives information on second order parameters	Only a couple of times per year on selected wafers

Table 2.3: Overview over the three different parameter classes

the stability of the electrical parameters at any time. For this purpose the Process Capability Indices (PCI) [60] are used to assess the process' ability to achieve yield. The two most common over-all metrics to assess a process's capability are C_p and C_{pk}. These indices were created in the Statistical Quality Control field. However, they are a useful metric even when evaluating compensation controllers.

The following calculations assume, that the parameter values measured obey a GAUSSIAN normal distribution function. A more detailed derivation of the underlying central limit theorem and the characteristics of the GAUSSIAN normal distribution and its parameters is given in Appendix A. C_p strictly evaluates a process's variability compared to the specification limits on that process:

$$C_p = \frac{USL - LSL}{6\sigma} \tag{2.10}$$

where USL,LSL are the upper and lower specification limits respectively and σ is the standard deviation of the normal distribution of parameter measurement values (e.g. electrical or geometrical data like threshold voltage or gate oxide thickness).

While C_{pk} also considers the mean of the process, i.e., how centered the process is within its specification limits:

$$C_{pk} = \min\left(C_{pL}, C_{pU}\right) \tag{2.11}$$

Where

$$
\begin{aligned}
C_{pU} &= \frac{|\bar{X} - USL|}{6\sigma} \\
C_{pL} &= \frac{|\bar{X} - LSL|}{6\sigma}
\end{aligned} \tag{2.12}
$$

with $\bar{X}$ as the mean value of the parameter measurements.

The desired values of C_p and C_{pk} are commonly said to be 2.0 for a 6σ process. Figure 2.13 shows a graphical representation of C_p and C_{pk} with a value of 2, along with a shift in the mean of 1.5σ yielding a C_{pk} of 1.5.

Note that when the mean shifts and the variance does not, as shown in Figure 2.13, the C_p value remains unchanged, as also shown in Figure 2.13. A C_{pk} of 2.0 means that only 2 parts per billion (ppb) are outside of the specification limits, while a value of 1.5 means that 3.4 parts per million (ppm) will be out of specification. These values translate to a yield of 99.9999998%

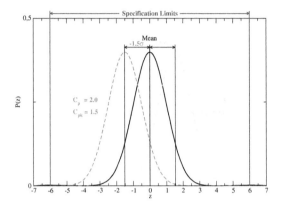

Figure 2.13: Example for parameter distributions and resulting C_p and C_{pk} indices

and 99.99966% respectively.

These parameters are calculated for every pass/fail parameter and if one of these parameters is below 1.0, improvement actions are undertaken (e.g. unit process specification limits are tightened) to improve the C_{pk} parameter well above 1.0.

The most common method for abnormality detection in semiconductor industry is statistical process control (SPC) [61]. SPC is an entire methodology, including addressing which actions to take upon detection of an abnormality and how to inform the operators of required actions. In traditional SPC, the expected variation is again assumed to be described by a normal distribution occurring around a mean value. In other words, the errors around the mean are assumed to be Identically, Independent Distributed Normal (IID Normal). This assumption leads to the normal distribution as proven in Appendix A. It is represented as:

$$y = \mu + \epsilon$$
$$\epsilon = IIDN(0, \sigma) \tag{2.13}$$

where y is the measured value, μ is the mean of the distribution for y, ϵ is the random error in measurement of y, and $IIDN(0, \sigma)$ is the normal distribution with mean 0 and standard deviation σ. Another form of representing the distribution of y is

$$y = N(\mu, \sigma) \tag{2.14}$$

with $N(\mu, \sigma)$ as the normal distribution function. In SPC an abnormality is assumed to be a shift in the mean of this distribution (μ), or a change in the standard variation of the normal distribution (σ). The abnormality detection technique is based on statistics and charting of the data. Different types of statistics have a different associated charting method. Thus, the specific fault detection techniques are usually called XYZ chart, with XYZ denoting the specific

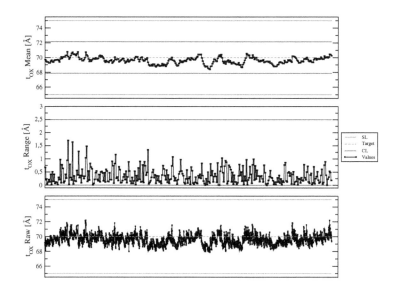

Figure 2.14: Example for a SPC chart showing a time series of thickness measurements of the gate oxide thickness

statistics used. Different techniques obviously test different hypothesies. Some test whether the mean (μ) has shifted, while others test whether the standard deviation (σ) has changed. Due to the statistics, it requires a much larger sample size to detect a change in the standard deviation than a change in the mean [62]. It is observed, that changes in the mean are more likely to occur. Consequently, charts to detect changes in the mean are much more common.

The most common SPC chart is the SHEWHART Chart, also known as an XBar-R (Average-Range) chart [63]. However in modern charting programs this naming convention is somewhat outdated (Average, Range, Raw Data and other statistical measures can be switched into the graph in every possible combination).

SPC control charts show drifts and trends in the monitored parameters. Additional control limits enable early warnings about possible instabilities far before the parameters go out-of-spec. An example of such an XBar-R control chart is shown in Figure 2.14. This chart shows one of the best controlled dimensions in semiconductor fabrication, the gate oxide thickness as a trend graph over around 3000 measurements.

A second example for an SPC chart is the variation of an electrical parameter. In Figure 2.15 the trend of the PMOS short channel threshold is shown. This data series includes approximately 10000 measurements over the time interval of a couple of weeks. Since this parameter is very sensitive to the PMOS channel doping and especially also sensitive to the thermal budget, it is much more difficult to control. The impact of corrective actions inside the fabrication and the

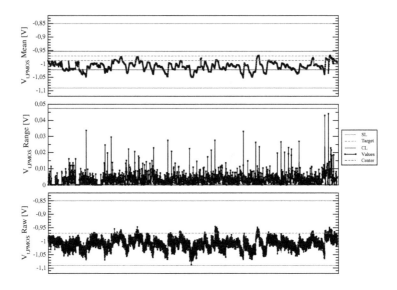

**Figure 2.15: Example for a SPC chart showing out-of-control events includ-
ing the impact of corrective actions for the threshold voltage
of a PMOS transistor**

influence of the control limits can be clearly seen in Figure 2.15. The control limits raise an early
warning flag and trigger corrective actions before the process gets out of control. The jumps in
the mean values indicated in Figure 2.15 indicate the impact of corrective actions in fabrication.

The second class of information parameters are measured and gathered as statistical data. However this data is neither subject to review during the pass/fail control mechanism nor subject
to SPC methods. Examples for such parameters are the effective channel mobility of a CMOS
transistor or its effective substrate doping.
The third class of parameters are characterization parameters which are difficult to obtain. For
this reason they cannot be monitored on a daily basis. They are updated on annual or bi-yearly
basis. Examples for such parameters are S-parameters or temperature coefficients.

2.7 Sort and Final Test

IC manufacturing processes tend to produce significant numbers of defective parts. Without
appropriate test procedures in place, the defective parts would find their way to customers and

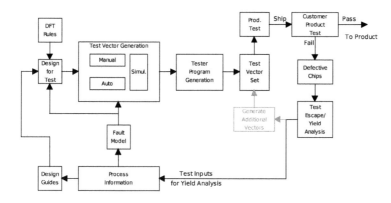

Figure 2.16: The test process

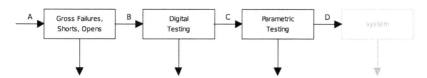

Figure 2.17: The three phases of IC testing

evidence themselves as poor quality. Furthermore, many ICs are used in security related systems (e.g. in a car) where it is definitely not acceptable, that only one defective part finds it's way into a system.

There are two instances of IC testing in a semiconductor manufacturing company. First there is sort which tests the IC on wafer level using probe cards similar to the procedure used at electrical test described in Section 2.6. The measurement equipment is about identical as described there, however, the architecture of the probe card is normally much more complicated, and the system is performing not only parametric but also digital pattern tests on the IC's. The metric commonly used to represent the quality of IC components is defect level (DL), also referred to as reject ratio. This is expressed by the ratio:

$$DL = \frac{N_{passed,bad}}{N_{passed}} \tag{2.15}$$

with $N_{passed,bad}$ as the number of *bad* parts which *test as good* and N_{passed} as the total number of parts passing the test. The defect level is typically expressed in ppm. (or defective parts per million, dpm).

Figure 2.16 indicates various aspects involved in test.

During the design phase, design for test (DFT) rules [64] and [65] are enforced and checked to ensure that tests of sufficiently high quality can be generated and applied. Figure 2.17 illustrates the flow of a typical chip test, in which there are three distinct phases.

Each phase rejects parts, and the quantity which the chip tests reject gives rise to the defect level discussed above.

2.8 Packaging

The role of packaging in semiconductor electronic applications is to protect and preserve the performance of the semiconductor device from electrical, mechanical, and chemical corruption or impairment. This role is becoming more and more important as well as difficult to execute as device performance, complexity, and functionality increase with each succeeding generation of technology.

The design of the package has a significant effect on the electrical behavior of an integrated circuit. The electrical representation of the package can be described in terms of a number of formats including resistance, inductance, and capacitance (RLC) [66]. A number of software tools are available to generate this RLC data and map it into a format for circuit simulation programs [67], [68].There are many programs, both commercial and university based, available for electrical modeling of packages and interconnects. Furthermore, the thermo-mechanical behavior of packages is getting more and more important. The change to area-array interconnect with high I/O counts and power dissipation has made thermal deformation an important concern for package reliability. In addition in smart power applications the self-heating of the big high-voltage output drivers can cause severe problems in the thermal design of the overall IC / Package / Environment system. Finally for high frequency applications in the GHz range, the RF/Mixed signal package modeling has to be carried out to predict the RF behavior of the overall system more accurately. In this application fields a careful design of the package by using simulation is mandatory. Since package modeling is a very complex issue for its own, and it is not scope of this work more detailed information may be found in, e.g., [69].

圖難於其易；
為大於其細。
天下難事必作於易；
天下大事必作於細。
是以聖人
終不為大，
故能成其大。
Lao-Tse (3rd or 4th century BC)

'... *Handle the difficult while it is still easy.*
Handle the big while it is still small.
Difficult tasks begin with what is easy.
Great accomplishments begin with what is small.
Therefore the wise
never strive for the great
and thus achieve greatness.'

Chapter 3

The TCAD Concept

3.1 Introduction and State-Of-The-Art

Multiple commercially and openly available TCAD simulation tools are available. In the following a short overview of the history of solutions for integrating them into a consistent work flow will be given. The following overview is far from being complete. However the main tools are reflected.

The history of commercial TCAD began with the formation of the company Technology Modeling Associates (TMA) in 1979. The software was a result of research performed at Stanford University under the guidance of Professors Dutton and Plummer. The most famous of the Stanford TCAD software programs are SUPREM and PISCES. SUPREM3 [70], [71] is a one-dimensional process simulator, while SUPREM4 [72], [73] can handle two dimensions. PISCES [74], [75] is the corresponding two-dimensional device simulator. These are general purpose simulators designed to work with fairly arbitrary semiconductor structures. TMA's versions of these programs were TSUPREM4 [76],[73] and MEDICI [77],[78]. Silvaco later licensed these programs from Stanford University too and offered a commercial alternative (ATHENA [79],[80] and ATLAS [81]). The third major TCAD vendor was Integrated Systems Engineering (ISE). Their equivalent product offerings were DIOS [82] and DESSIS [83].

TMA was later acquired by Avant! which was then acquired by Synopsys. Recently Synopsys acquired their main competitor ISE, which leaves only Synopsys and Silvaco as the main competitors in the market. Synopsys owns now more than 80% of the market share, which results in a market situation close to a monopoly.

3.1.1 Process Simulation

Process flow simulators are tools which simulate the full semiconductor manufacturing flow in a certain level of detail. For special applications and special tools (like lithography) the level of detail of the models implemented in generic process simulators is not sufficient to cover effects like, e.g., simulation of proximity effects in lithography or detailed etch sidewall shapes in plasma etching. However for the routine task of, e.g. generating a accurate representation of a semiconductor device suitable for device simulation the available process simulators are sufficient.

3.1.1.1 Generic Process Simulators

1. SUPREM3 the "mother of all process simulators" is now completely outdated. It is a one-dimensional process simulator incorporating already sophisticated models like diffusion in polycrystalline layers and point-defect diffusion.

2. SUPREM4 is the basis for the two commercial tools TSUPREM4 and ATHENA. Developed at Standford University in the group of Prof. Dutton, it was the first consistent approach to simulate the physical behavior of dopants in a layered two-dimensional cut through a semiconductor wafer during semiconductor processing.

3. FEDSS is based on the finite element method too. It was developed inside IBM [84] for generating suitable device structures for the device simulator FIELDAY [85]. It was a fully featured process simulator. However the level of detail of especially the diffusion models was much less sophisticated than that inside of DIOS or SUPREM4.

4. PROPHET is comparable to FEDSS and was developed in AT&T [86]. Where the main focus was the simulation of BiCMOS technology.

5. DIOS is a two-dimensional process simulator developed initially by the Swiss company ISE. Among the strengths is the adaptive meshing for structures with difficult aspect ratios (e.g. smart power devices) and the big variety of implemented models. Drawbacks are the inconsistencies in the different granularities of the models. For instance in the diffusion models the equilibrium diffusion parameters cannot be used as a basis for the point-defect diffusion models.

6. TSUPREM4 is a two-dimensional process simulator descendant from the Stanford process simulator SUPREM4. Strengths are the stable and clear programming interface and the consistent set of simulation model parameters. A weakness is the clumsy meshing and the need for setting up an appropriate initial mesh. Especially for automated process split simulations it often breaks down because of an inappropriate initial mesh.

7. TAURUS-PROCESS was a very ambitious approach to implement a process simulator based mainly on the level-set algorithm [87], [88]. The main problem of three-dimensional process simulation, to cope with moving internal and external three-dimensional boundaries and especially topological changes, when certain parts or entire layers are "consumed" during a semiconductor manufacturing process step, was shifted from surface meshing to bulk meshing. However the development failed to provide a stable three-dimensional process simulator yet.

8. ATHENA is the third commercial process simulation code and an additional descendant from SUPREM4. It is nearly identical to the TSUPREM4 simulator.

9. FLOOPS is an object oriented level-set based process simulator from the University of Florida [89]. It uses the algorithmic language ALAGATOR to enable the implementation of new models into the internal discretization scheme. Recently Synopsys is developing a

commercial version of FLOOPS for three-dimensional process simulation and as a successor of DIOS

3.1.1.2 Equipment Simulators

Equipment simulation is still an area which is strongly under development. The first stable equipment simulators were lithography simulators which try to analyse the complex sequence of resist spin-on, illumination, development and strip with finite-element methods. Equipment simulators for etching and deposition are still strongly limited to certain manufacturing tool-sets (like those of Applied Materials) [90],[91],[92] and are normally not able to cover the full range of machine parameters which can be tuned at a certain equipment. There are also no commercially available general equipment simulators on the market.

1. ILLUM2D/3D is a tool developed by the Institute for Microelectronics at the Technical University of Vienna, which models the full lithography process flow. It is well suited to capture the physics of the process module like illumination. However the chemical effects during post-exposure bake and development are not well covered by the tool.

2. SPLAT/SAMPLE2D/3D are tools for aerial image simulation and two-,three-dimensional lithography and topology simulation developed by the University of Berkeley in the group of Prof. Neureuther [93].

3. PROLITH is a fully featured lithography simulator now distributed by KLA Tencor [94]. It offers two-dimensional and three-dimensional functionalities and is a standard tool used by the lithography groups worldwide.

4. SOLID E is the competitor to PROLITH and offers a comparable set of features for three-dimensional lithography simulation [95].

5. ACES (Anisotropic Crystalline Etch Simulation) is a tree-dimensional etch simulator using a continuous cellular automata (CA) model and a dynamic structure update method [96]. The program can simulate silicon etching with different surface orientations in selected etchants with variable etch rate ratios. It can receive two-dimensional mask designs in common mask formats (including CIF, GDSII, BMP) and generate three-dimensional profiles in standard solid-modeling languages.

3.1.2 Device Simulators

Device simulators are the counterparts for the process simulators shown above. However historically device simulation was done much earlier than process simulation. Based on assumptions on the input structure of semiconductor devices pioneering work on device simulation was carried out at ATT [97] and IBM [98] leading to major university efforts such as TU Vienna [99] and Stanford [100], finally culminating in a rapid growth of TCAD vendors and development of commercial platforms that support a broad and heterogeneous set of users.

3.1.2.1 Generic Device Simulators

1. PISCES 2ET is a dual energy transport (for carrier temperatures and lattice thermal diffusion) semiconductor device simulator. Some advanced features are the simulation of the carrier and lattice temperatures, and heterostructures in compound semiconductors. Hence, various non-stationary phenomena such as hot carrier effects and velocity overshoot can be analyzed using this program. The electrical behavior of optoelectronic devices can also be simulated with reasonable accuracy. Most of the material parameters have been calibrated and thoroughly surveyed with the help of industry.

2. FIELDAY [85] is a simulator for devices of arbitrary shape in one- up to three-dimensions. The models are especially tailored for the analysis of bipolar devices but field effect transistors can be modeled too. The complementing program FEDSS is used for the generation of input structures.

3. PADRE [101] is comparable to FIELDAY and an internal development of AT&T. It is a moment based device simulator.

4. MINIMOS-NT is a general-purpose semiconductor device simulator providing steady-state, transient, and small-signal analysis of arbitrary two- and three-dimensional device structures [102]. It was recently compared to devices simulated with DESSIS, and it yields the same quality of results as the commercial tool.

5. MEDICI is the counterpart of TSUPREM4 on the device simulation side. It is a pretty stable hence fairly old device simulator which can deal with a variety of physical effects in two-dimensional semiconductor structures. The code was licensed from Stanford University and bases entirely on the PISCES code.

6. DESSIS is a very sophisticated device simulator which deals with two- and three-dimensional device structures. It has a fairly similar feature list compared to MEDICI. However, it is much more stable and based on newer source code (C++ instead of FORTRAN) than MEDICI [103].

7. ATLAS is nearly identical to MEDICI in terms of features [81].

8. FLOODS is the counterpart to FLOOPS on the device simulation side [89].

3.1.2.2 Specialized Device Simulators

Specialized device simulators work on a "device template" input structure. These templates are predefined or even hardcoded in the device simulator. The simulator assumes a certain type of device (e.g. a MOSFET for MINIMOS [104] and PISCES [105] or a bipolar transistor for BIPOLE) and takes values for predefined dimensions (e.g. gate oxide thickness, gate width etc.) and doping profiles (e.g. gate channel doping profile assumed as GAUSSIAN distribution) of the semiconductor device.

1. MINIMOS is the predecessor of MINIMOS-NT. It is the famous MOSFET device simulator which worked on an orthogonal grid generated internally [106]. The simulator was a major breakthrough in the theoretical investigations of semiconductor devices, because a lot of physical effects were in reach of detailed analysis for the first time.

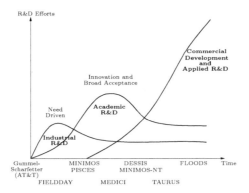

Figure 3.1: Schematic time-line of TCAD R&D for device analysis

2. PISCES is a simulator comparable to MINIMOS, developed by Stanford University [100].

3. SEQUOIA [107] device designer is a simulator comparable to MINIMOS, developed by Sequoia Design Systems.

4. BIPOLE [108],[109] is a simulator for device simulation of bipolar transistors. The core code deals with the analysis of a one-dimensional cut through the emitter/base/collector region of an integrated bipolar transistor device. It adds additional parasitic elements into the analysis like the collector resistance by analysing the geometric dimensions and sheet resistances of the device.

The tools that define the TCAD field - process, device and circuit modeling - have evolved steadily over the past three decades, moving from research prototypes (both in industry and academia) towards robust workhorses that support both research and manufacturing applications. Figure 3.1 shows a schematic time line of evolution for process and device simulation. It is obvious that the development efforts of the commercial vendors have been steadily increasing since the 80's.

3.1.3   TCAD Work Flow Environment Software ("Workbenches")

Every commercial TCAD vendor is including an environment software for automated or at least semi-automated setup of the TCAD work flow shown in Figure 3.2 in Section 3.2. The three workbench software tools are GENESISe from the former *ISE AG*, WORKBENCH from *Synopsys Inc.* and VIRTUAL WAFER FAB from *Silvaco Int.*. Another free TCAD environment software is VISTA [110],[111] from the Institute for Microelectronics, TU Vienna. However this software is not under active development at present.

All four environments are fairly similar in their architecture. They offer macros to schedule and generate Design of Experiments (DOE) [112] simulation runs. For this purpose the simulator command files can be parametrized by a special syntax which identifies the position of a dedicated parameter in the command file. The types of parameters supported are shown in Figure 3.6 in Section 3.4. Furthermore, the extraction of dedicated results from process or device simulation (values given in certain simulation logs or result files) is supported by defining regular expressions or output templates. However all of these environments do NOT support interfaces from semiconductor manufacturing equipment or metrology tools. They offer a more or less well integrated work flow for performing simulations, but the support for interfaces from and to the simulation environment is very little or even not implemented. In the following sections a concept is layed out, how to set up such interfaces in a most effective and stable manner.

3.2   Overview

This section concentrates on the overall structure of the TCAD system as implemented during this work. The simulation work flow is outlined and certain specialities of this approach are shown. This work flow refers to the TCAD software suite of former ISE AG (now acquired by Synopsys) including the main simulators DIOS [82] (process simulator) and DESSIS [83] (device simulator). Since this work flow is only one of multiple possible implementations, some parts of this section cannot be generalized or applied to other TCAD installations (like TSUPREM4 and MEDICI installations) but care has been taken to show the TCAD-work flow as general as possible.
The different levels of functionality of the TCAD system are shown in Figure 3.2.

The purpose of the process simulation is to provide the structural information of the device under scope, consisting of the boundary including the composition of the different materials involved (e.g. polycrystalline silicon, single crystalline silicon, silicon dioxide, metals etc.). In addition, the doping concentration inside the silicon has to be available. The process simulation takes the photo mask information and the process flow to model the evolution of above mentioned information (boundary and dopant) over the multiple steps of the process.
The mesh for solving the partial-differential equations typical for the physical and chemical processes occurring during processing is normally of unstructured type, to model the steep gradients of the doping distributions with good accuracy, but with a low number of mesh points where the physical fields (doping concentration, point defect concentrations etc.) are not varying much. A detailed description of a process simulator can be found in, e.g., [113],[114],[115]
The boundary and dopant information is then used as an input to describe the electrical behaviour of the device under scope by calculating the potential distribution and the carrier transport phenomena (current concentrations etc.) via solving the PDEs describing their physics. A detailed description of the underlying principles of a device simulator can be found in, e.g., [116],[117],[106]
Since the requirements on meshes for process and device simulations, respectively, are very different, a re-meshing step is necessary to minimize the numerical error, and the number of mesh points necessary for a certain accuracy of the solution. This re-meshing is normally based on the gradient or difference refinement criteria. In some cases this approach is not sufficient to get a good mesh. The inversion region of a MOSFET channel is a good example for the problems gradient refinement criteria are facing. However in recent investigation approaches are outlined to overcome or, at least, to tackle these limitations [118],[119].

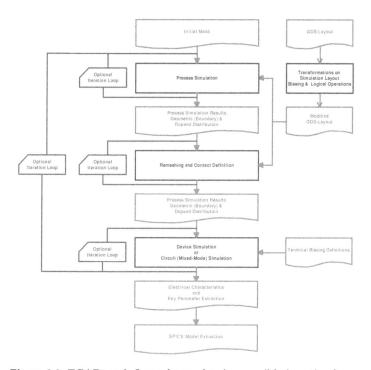

Figure 3.2: TCAD work flow scheme showing possible iteration loops

3.3 TCAD Input

To start with the process simulation block some information has to be prepared in a certain level of detail.

3.3.1 Layout

As in real semiconductor processing the layout of the device or integrated circuit is a substantial input to the fabrication flow. It defines the lateral composition of the circuitry and is the main variable input to the semiconductor fabrication line [1]. The layout consists of the combination of a set of different mask levels. Each mask level is defining a certain functional block during processing. As one example, the gate level defines the sizes and orientation of any CMOS gate inside the integrated circuit. As another example, the Boron source/drain masks define the areas

---

[1] except the process technology itself which could be varied based on the needs of the product, however in a standard semiconductor fabrication the number of different process technologies is far less than the number of integrated circuits on ONE technology platform

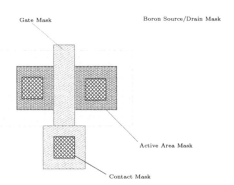

Figure 3.3: Simple layout example of a PMOS transistor

where the Boron for the PMOS source/drain regions has to be implanted. The combination of these mask levels characterizes the overall structure of certain devices. A simple example of this concept is shown in Figure 3.3.

It can be seen from this figure that the source/drain implant is drawn over the gate layer too. However since in the fabrication process the gate layer is masking the implantation, the real source/drain diffusion is only the logical NOR operation of the two layers.

3.3.2 Mask Bias

Mask bias is a post processing step during the fabrication of the chrome reticle masks for lithography. Especial the patterning process is subject to process related variations which have to be compensated with lithography. A good example is the somewhat outdated [2] process module LOCOS (LoCal Oxidation of Silicon) [120] isolation. Figure 3.4 shows the detailed process step chain of this module.

Figure 3.4(a) shows the initial stack of the LOCOS sequence consisting of the single crystalline silicon substrate, the so called pad oxide consisting of Silicon Dioxide as a stress relief layer and Silicon Nitride which acts as oxidation suppression mask for the oxidation of silicon. Figure 3.4(b) shows the situation after spin-on of the photo resist. On top of the structure the chrome reticle dimension for the subsequent illumination step is shown. Furthermore, the initial dimension of the mask data is shown. Since the drawn layout dimension should define the final size of the area of the active region (the region not covered by field oxide), the real chrome area has to be bigger (biased) in size. A detailed outline of the algorithm for proper biasing of the two-dimensional layout structures is given in Appendix C. Figure 3.4(c) shows the situation after illumination, development and hard bake of the photo resist. A couple of effects, like under or overexposure, resist shrink during post exposure bake etc., may lead to differences in the width of the photo resist structure and the initial reticle dimension. These differences are the first contribution to the CD (critical dimension) difference between the initial reticle size and

---

[2]the module is the standard approach for the transistor isolation for technology nodes $\geq$ 250nm

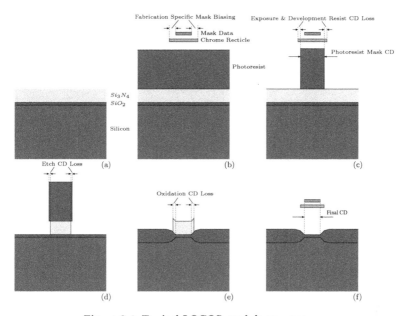

Figure 3.4: Typical **LOCOS** module sequence

the final size of the structure of silicon. Figure 3.4(d) shows the CD loss because of a small undercut of the Silicon Nitride during layer etch. Additional CD loss may occur if the etching chemistry consumes small parts of the capping photo resist layer. Figure 3.4(e) shows the CD loss due to encroachment of the silicon nitride capping layer during oxidation of the silicon. Finally Figure 3.4(f) gives the situation after stripping of the capping nitride layer. If the initial reticle mask bias compensates for the above mentioned effects during the steps (c) through (e) in Figure 3.4, the final CD matches exactly the initially drawn layout dimension. Since the processing effects depend critically on the details of the manufacturing process, this biasing is called fabrication specific.

Another important convolution of the initial mask input to process simulation are the mask proximity effects due to the diffraction effects during photo resist exposure. The underlying physics was described in Section 2.4 in detail. A simulation taking into account these effects is outlined in Section 6.3. An example of the magnitude of these effects for a typical 350nm node mask illuminated with i-line lithography is shown in Figure 3.5.

Figure 3.5(a) shows the initial layer data for some levels during fabrication of a EEPROM cell [11],[122] (Floating Gate, Gate Poly, Contact and Metal 1). Figure 3.5(b) through Figure 3.5(e) show the intensity distribution during resist illumination. Figure 3.5(f) shows the extracted iso contours (at same level of intensity) of the 4 layers demonstrating the above mentioned proximity effects during lithography.

By applying these simulations the resulting proximity corrected contours can be used as an in-

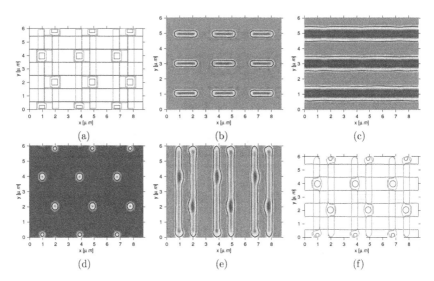

Figure 3.5: Initial layout of a part of an **EEPROM** cell (a), contour plots of intensity distribution during the illumination of photo resist at different levels (b)-(e) and resulting resist contours after development (f).

put to the process simulation in the usual CIF or GDSII formats. Therefore, the layout may be preprocessed once to reflect the real mask shapes more closely. To simplify the resulting contour polygons a polygon point reduction algorithm, the minmax-method [123] has been used. An example of the results of such a preprocessing of the mask layout can be found in Section 6.3

3.3.3 Process Conditions

In Chapter 2 it has been outlined, that the clean room production flow is a fairly linear flow but of high complexity with respect to the production path in the fabrication. Since this flow has to be documented extremely well to prevent misprocessing, a so called MES (Manufacturing Execution System) is used to control the wafers during their full processing flow. Controlling means tracking of the current position in the process flow and defining the correct machine recipes at the correct positions of the flow. This process flow (to name it short) is a list of single process steps which determine WHEN and HOW the wafer surface has to be modified by implantation, etching, deposition, and masking to form an integrated circuit at the end.

There are two classes of process simulation steps. On one side the "physical" process steps ion implantation, oxidation and diffusion where the machine parameters (recipes) are inputs to the process simulation also. In contrast to this class there are the "chemical" process steps etching

Step	Low Detail		Medium Detail		High Detail	
diffusion	temp.	pressure	temp.	pressure	temp.	pressure
&	time	flow rates	time	flow rates	time	flow rates
oxidation			temp. ramps		temp. ramps	gas ramps
	of main diff. step		*of all diffusion steps*		*of all diffusion steps*	
ion	dose	energy	dose	energy	dose	energy
implantation	tilt		tilt	revolving	tilt	revolving
					beam div.	ioniz. level

Table 3.1: Examples for current level of details in the description of semiconductor process steps

and deposition, where currently there are no stable, reliable and sufficiently fast models available to model these steps with their process recipes (pressure, temperature over time) rigorously. These steps are modeled via simple geometric "emulation" of the outcoming topology change after the run recipe. Therefore, the calibration of these steps is very important. Lithography is in part "physical" (illumination) and in part "chemical" (PAC reaction and development), and normally it its treated as deposition and etching in terms of geometrical modeling. In terms of illumination it has been shown in the previous section, how to care about these effects.

To finally get the deserved simulation results by process simulation with appropriate accuracy, it is mandatory to mimic the real processing with a high level of detail provided by the physical models of the process simulator. A good example of such an approach is the implementation of a diffusion and oxidation recipe in a process simulator as shown in Section 6.1. In this example every single change in temperature (temperature ramps) or pressure (pressure ramps) or gas ambient change (oxidizing ambient instead of inert ambient) above a certain temperature where the diffusion models of the process simulator are calibrated and valid, was defined as a single command for the process simulator. The final simulation program consists of a sequence of different process conditions of a certain time. In the past this concept was not employed because of the big penalty in simulation time. In contrast only the "main thermal step" was defined as a process simulation command. The new concept has the big advantage of being not prone to the judgement of the TCAD engineer, which of the numerous single diffusion steps (between 8 and 30) is now the most dominant and important one.

In Table 3.1 a comparison of the current levels of details in the process flow information according to the current status of the physical models of the available process simulators is given. The process steps deposition, etching, and lithography were left out on purpose since these steps are normally not modeled via equipment level simulation (input parameters to the simulation are the process recipes) but via geometrical approximation as outlined above.

3.4 Process Simulation

Process simulation models every process step of a silicon wafer from start of processing until electrical wafer acceptance test. The different process steps possible were outlined in Section 2.5. As mentioned in the previous section, the level of detail of input parameters to the simulator is

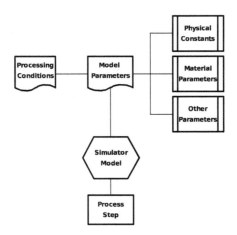

Figure 3.6: Parameter classes for process simulation

heavily depending on the complexity of the physical model implemented in the process simulator. These input parameters are varying, depending on the type of process step. An example of such parameters is given in Table 3.1.

The process simulation flow consists of a sequence of single commands for the process simulator, naming the type of the process steps (diffusion, etching, etc.) and its process conditions in terms of parameters (time, temperature, pressure, etc.). A detailed review of the syntax of such a process flow is given in Chapter 5.

A second class of parameters are the *model parameters* which are the result of a careful calibration of the models for a certain range of input parameters (like e.g.temperature). These parameters describe the underlying physics of the process simulator model. Figure 3.6 shows this situation.

There are several types of applications for process simulation in the semiconductor manufacturing industry and research.

First, (especially during the process flow development phase) single process steps are simulated and optimized for a certain targeted feature like oxide distance or well depth. A good example of such optimizations is the transfer of furnace recipes between different wafer sizes given as a detailed example in Section 6.1. In these situations one is mainly interested in optimizing the main step of a diffusion recipe to gain identical thermal budgets for both recipes thus obtaining the same diffusion profile on different wafer topologies. The algorithm is shown in Figure 3.7(a).

Second, during integration of additional modules in an already available process flow, process simulation is used to identify key influences of the module on the overall process flow. The term module means a combination of process steps which form a unit to make a certain feature of the process technology like the isolation between semiconductor devices (LOCOS for older technologies and shallow trench isolation for newer technologies). A good example is, e.g., added thermal

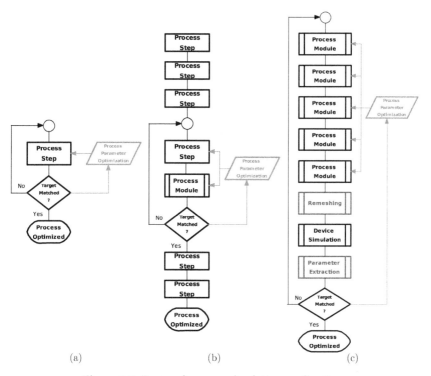

(a) (b) (c)

Figure 3.7: Types of process simulation applications

budget because of additional diffusion steps inside the new module. This additional thermal budget must be minimized to gain the same features of process steps preceding the module as given by the process flow without the additional module (e.g. the doping profile of a preceding threshold adjust implant is broadened by the additional thermal budget, thus the implant dose has to be increased and/or the implant energy lowered to maintain the same surface concentration of the threshold adjust implant). The algorithm is shown in Figure 3.7(b).

Third, an already installed process flow is simulated to generate the electrical parameters of semiconductor devices by a process simulation followed by a subsequent device simulation. This type is mainly used for optimizing the process technology in terms of parametric performance. Furthermore, in terms of process complexity (cost) simulations are carried out to optimize an already implemented process technology further. In addition the statistical sensitivity of device parameters (like threshold voltage or on-resistance) on certain process flow parameters like implant energy or dose of a certain implant can be calculated. The algorithm for this strategy is shown in Figure 3.7(c)

3.5 Re-Meshing and Boundary Processing

The requirements on grids of process- and device simulators are very different. On one hand the process simulation grid has to follow steep gradients in the doping profile and must resolve internal interfaces like SiO_2/Si with an accuracy sufficient to model segregation and dopant transport across interfaces [124],[125]. Furthermore, it must be able to adapt to changes in the boundary occurring during process steps like oxidation or etching. These changes can lead even to topology changes (like etching a hole through one layer converting one segment into two separated segments) [87],[88],[126]. On the other hand the device simulation grid must resolve mainly the fields of physical quantities occurring during a device simulation, like carrier concentrations or electrostatic potential. Thus the grid obtained by the process simulation is normally not suitable to get sufficiently accurate results for device simulations.

Therefore, re-meshing of the resulting structure after process simulation is mandatory. Re-meshing is a very sensitive process, since it includes interpolation of the dopant distributions on a new grid [127],[128]. Care has to be taken, not to increase the errors in the mesh representation of the dopant field [129]. Especially the boundaries where currents occur during the device operation have to represented appropriately by the mesh. A proper grid for carrier transport simulation should follow the current flows during device operation (like a refined channel in a MOS structure for simulation of the transfer characteristics of a MOS device).

There are three strategies to generate a suitable grid. A straight forward and very stable method is the generation of the grid based on dopant gradient criteria. For certain applications these methods have big disadvantages, because they do not generate a high resolution in areas where necessary. For instance the channel, of a MOS transistor shows no steep dopant gradient and is thus not well resolved by a standard gradient criteria based grid. One workaround is to define a dedicated refinement region with a finer resolution in these areas. However, this approach is not suitable for automatic re-meshing of different device types.

A newer approach is the generation of boundary conforming meshes [130]. In this approach the boundaries of a certain material segment are the starting conditions for a mesh generation by offsetting mesh lines from the boundary by a certain distance, which increases as the mesh lines

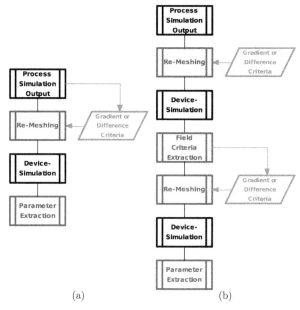

(a)   (b)

**Figure 3.8: Strategies for re-meshing structures generated by process sim-
ulation. (a) Straight forward single step method (b) Two-step
method**

propagate into the material segment. This method yields excellent results for resolving critical
channel areas as discussed above. However, it generates unwanted mesh points in areas which
are not related to the active region of devices. Therefore to suppress this generation, a lot of
user interaction has to be performed, do define the interesting segments in the structure.

A more general and automatic approach is a two step strategy to generate a suitable device
simulation grid. First a coarse grid is used with the device simulator to obtain a coarse representation of the physical fields in the device. Based on this solution a physical field is chosen as
the refinement criteria for a second iteration of the re-meshing process. For a CMOS device an
appropriate field would give the carrier density in the device. One Drawback of this approach
is the possibility of big errors in the initial solution of the physical fields, which could lead to
convergence problems in the second iteration of the re-meshing. Generally the two step strategy
demands more calculation time. For big two-dimensional grids or three-dimensional grids this
method can, however, lead to much faster device simulations because of improved convergence.
Some problems cannot even be solved with a simple one-step approach. Figure 3.8 shows the
algorithm of both methods including the possible re-meshing criterion.

A simple example of the refinement of a process simulation mesh of a CMOS transistor which
has to be simulated for drain to substrate breakdown is shown in Figure 3.9.

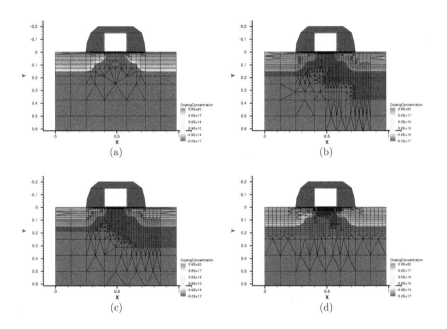

Figure 3.9: Comparison of automatically and manually refined meshes (a) Initial coarse grid (b) 1st iteration (c) 2nd iteration (d) manually refined grid

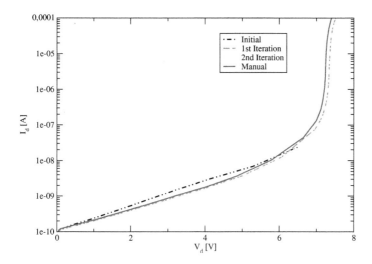

Figure 3.10: Comparison of the different meshes in terms of device simulation results

In Figure 3.9(a) a very coarse initial grid interpolated from the process simulation grid is shown. This grid is used for a first analysis of the junction breakdown between the drain and the substrate of the NMOS transistor. The resulting electric field inside the device extracted when the reverse current between drain an substrate reaches a certain level is used with gradient refinement criteria for the first iteration shown in Figure 3.9(b). The results of a device simulation with this grid are used for a 2nd iteration with the same criteria. The resulting mesh is shown in Figure 3.9(c). It can be seen clearly from this figure, that the fine mesh follows the field distribution in the breakdown situation very smoothly. As a comparison a mesh which is created by manually placed refinement boxes is shown in Figure 3.9(d). The electrical characteristics of the reverse biased drain/substrate diode for these different grids can be seen in Figure 3.10.

The initial grid gives a moderately inaccurate result (the simulation was stopped when the ionization integral inside the structure reached unity to speed up the simulation). The iterative and the manually optimized meshes give nearly identical results. However, to set up such a mesh manually human interaction is necessary, which is not suitable for automatic simulation flows. Therefore the approach to generate the mesh refinement based on initial device simulations is the method of choice for a stable automated mesh generation.

Drawbacks of this method can be an increased mesh node count, because the refinement is not restricted to certain parts of the device.

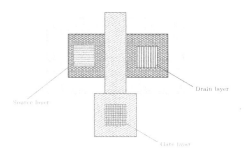

Figure 3.11: Simple layout example for contact naming of a MOS transistor

3.6 Contact Definition

Before a device simulation can be started, the contacts of the device have to be defined, since these contacts represent the geometric position of the electrical boundary conditions. The simulator must know a unique name of each contact to identify it during the device simulation in order to set the boundary conditions (potential, current, parasitic values, work function etc.) appropriately.

To enable an automatic generation of these contact names, the names have to be defined in the device layout. These regions have to be projected on the final two- or three-dimensional structure in a post processing step of the results of the process-simulation. If the contact naming follows a certain nomenclature, standard device simulator templates for certain semiconductor devices can be used automatically. If, for instance, every source, drain and gate of a CMOS transistor is named *source*, *drain* and *gate* consistently in every layout, a standard template for the device simulator command file can be used to extract electrical parameters, like threshold, saturation current, etc. This naming is performed by using the CONTACT layer data, to identify the different contacts. For this purpose some selected contact structures are copied to the corresponding contact naming layers like **source**, **drain**, **gate** etc.

An example of such a naming approach is shown in Figure 3.11.

3.7 Device Simulation

Finally, the data structure is ready for processing with device simulation. The main inputs for device simulation are:

1. Doping concentration of different doping species (e.g. Arsen, Phosphorus, Antimony, Boron etc.) on a mesh.

2. Structural information about the shape of the region which has to be evaluated with device simulation. This information includes material types of layers, topological variation of layers and the detailed surface and interface shapes of the materials present in the structure.

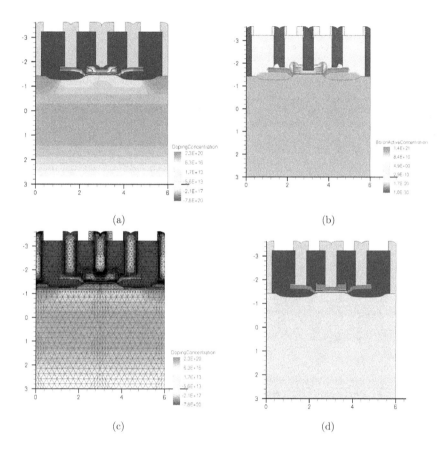

Figure 3.12: Example for a typical input structure for device simulation
comprising of doping concentration (a) including the different
doping species e.g. Boron (b) on a mesh and the topological
structure and contact definition (d).

3. Named contacts as source for adjustable boundary conditions of the device simulation.

An example for a typical input structure for device simulation is shown in Figure 3.12. The three
main input classes (doping, structural and contact information) can be clearly seen.

The semiconductor device simulators are fairly similar in their solution approach. They all solve a
system of partial differential equations describing the potential distribution and carrier transport

in a doped semiconducting material. The standard semi-classical transport theory is based on the BOLTZMANN equation [131],[132]

$$\left(\frac{\partial}{\partial t} + \vec{F}\nabla_{\vec{p}} + \frac{\vec{p}}{m}\nabla_{\vec{r}} \right) f(\vec{r}, \vec{p}, t) = \left(\frac{\partial f}{\partial t} \right)_{coll} \tag{3.1}$$

where $\vec{r}$ is the position, $\vec{p}$ is the impulse, $\vec{F}$ is the electric field vector and $f(\vec{r}, \vec{p}, t)$ is the distribution function. In the simplest approach for solving this equation the collision term on the right hand side of (3.1) is substituted with a *phenomenological* term

$$\left(\frac{\partial f}{\partial t} \right)_{coll} = \frac{f_{eq} - f(\vec{r}, \vec{p}, t)}{\tau} \tag{3.2}$$

where f_{eq} indicates the (local) equilibrium distribution function, and τ is a microscopic relaxation time. It is very useful to express the distribution function in terms of velocity, rather than impulse, since it will be easier to calculate electrical currents. In equilibrium one may use the MAXWELL-BOLTZMANN distribution function

$$f_{eq}(\vec{r}, \vec{v}) = n(\vec{r}) \left(\frac{2\pi k_B T_0}{m^*} \right)^{-\frac{3}{2}} e^{-\frac{m^*|\vec{v}|^2}{2k_B T_0}} \tag{3.3}$$

where $n(\vec{r})$ is the carrier density, T_0 is the lattice temperature and m^* is the effective mass. The use of (3.3) for semiconductors is justified in equilibrium as long as degeneracy is not present. the carrier density $n(\vec{r})$ is directly related to the distribution function according to

$$n(\vec{r}) = \int d\vec{v} f(\vec{r}, \vec{v}) \tag{3.4}$$

which is of general applicability. The significance of the momentum relaxation time can be understood if the electric field is switched off instantaneously and a space-independent distribution is considered. The resulting BOLTZMANN equation is then

$$\frac{\partial f}{\partial t} = \frac{f_{eq} - f}{\tau} \tag{3.5}$$

which shows that the ralaxation time is a characteristic decay constant for the return to the equilibrium state.

The often used drift-diffusion current equations

$$\begin{aligned} J_n &= qn(x)\mu_n F(x) + qD_n \frac{dn}{dx} \\ J_p &= qp(x)\mu_p F(x) - qD_p \frac{dp}{dx} \end{aligned} \tag{3.6}$$

can be easily derived directly from the BOLTZMANN equation as outlined in Appendix D. All device simulators use the drift-diffusion approach as the simplest model to cover the transport effects inside the semiconductor material.

3.8   Electrical Key-Parameter Extraction

To evaluate the electrical characteristics of a semiconductor device, the results of device simulation can be evaluated. Typical key parameters are the linear or saturation threshold V_t [133], the transconductance g_m, or the saturation current I_s of a CMOS transistor. For bipolar transistors the common emitter current gain β, the early voltage V_{AF}, or the collector resistance R_C are examples for common key parameters.

Typically, TCAD vendors offer standard algorithms for extracting these parameters with post processing tools or scripts after the device simulation (e.g. Inspect [134] or Tonyplot [10] are such tools for showing electrical characteristics and extracting key parameters). The main basis for the definition of key electrical parameters are the SPICE models like BSIM3[3] [135],[136], however for manufacturing control these parameters are too cumbersome to obtain and the number of measurements is too high for gaining enough throughput in parameter test. Therefore fast extraction algorithms working with a few measurement points (e.g. threshold voltage is often extracted only with 5 current-voltage measurements) are thoroughly used. However, the algorithms may vary strongly between different semiconductor companies, since even a simple parameter like the CMOS threshold voltage may be measured in a lot of different ways. Therefore, a careful setup of these parameter extraction algorithms in simulation is very important to enable calibration of simulation against measurements and to benefit from the enormous database of electrical test data for calibration of simulation.

A method for setting up these algorithms in a very efficient way is shown in Chapter 4.

3.9   Optimization and Inverse Modeling

A newer trend in TCAD simulation is the use of automatic optimization and inverse modeling [137],[138],[139].

A special example for optimization of the manufacturing process flow was already given in Section 3.4. Additional application of optimization are the optimization of device layout and of device simulation parameter classes like outlined in Figure 3.6 for semiconductor process flow optimization. The typical optimization approach of the commercial TCAD vendors is the generation of response surfaces [140] via definition of DOEs and the analytical calculation of the optimization minimum from the generated model [112]. This type of optimization environment is offered inside the graphical environment of the user interfaces of the TCAD systems as outlined in Section 3.1. The advantage of this approach is the stability of the optimization. Even if one or two simulations fail (e.g. because of mesh stability or accuracy problems), a good result of the optimization can often be gained. However, a major drawback of this approach is, that the input parameter interval cannot be set very broadly, because of the computational costs (even if DOE methods are applied). Therefore it often happens (as also with experiments in fabrication), that the final optimum is outside the defined input parameter limits.

The other approach is the use of real multidimensional optimization algorithms like *downhill simplex*, *direction set* methods of the class without calculating derivatives or *conjugate gradient*, *quasi-Newton* and *variable metric* of the class of methods calculating first-order derivatives, and,

---

[3]The newer CMOS model BSIM4 has many advantages for modeling deep sub-micron devices, but the definition of key electrical parameters like threshold voltage gets more and more ambiguous, since the newer models tend to use the compact modeling and not the physical modeling approach.

finally, *simulated annealing* and *genetic algorithm* methods which form a class of their own in terms of mathematical tools used. Details of these methods can be found in, e.g., [141],[142],[143]. A special application of optimization is inverse modeling [142],[144],[145]. Basically it is identical to optimization, however the target is a different one. In optimization the scope is the optimization of the overall system to gain a more efficient manufacturing method. Inverse modeling aims not to gain an optimized system at the end, but to get information not accessible to forward analysis. Inverse modeling defines an analytical or numerical model with a certain set of input parameters and compares this model with a desired result (e.g. a measurement). A score function of the type as shown in (6.1) in Section 6.1 may be used for such a setup. After finding a global minimum of the score function the model is considered to be reflecting the physical parameters. A good example is the class of convolution problems in metrology methods. For instance, SIMS or SRP measurements are convoluted with the internal point response functions of the measurement systems. These point response functions are defined by the physical effects occurring during measurement (e.g. ion mixing during SIMS sputtering [146] or carrier spilling [147],[148] during SRP measurement) and can be modeled with a simulator. The "real" doping profile may be obtained by inverse modeling as above mentioned.

Chapter 4

Integration between Semiconductor Fabrication and TCAD

4.1 Introduction

After the general overview about the environments of semiconductor integrated circuit fabrication and the corresponding TCAD simulation environment, the following chapter deals with the structure and approach of the new concept. This concept is connecting both worlds in an automated way. Furthermore, the painstaking process of getting data from one world into another is dramatically reduced and the overall productivity of TCAD simulation is significantly improved.

The next section lays out the different interfaces between "reality" and "simulation" in a structured way. The basic information paths are identified and analyzed. This information is important to specify the border conditions for the different converters shown in the next chapter. In addition, a good picture of the different levels of information input into the simulation framework is given.

4.2 Interfaces

The different interfaces identified can be seen in Figure 4.1. The shaded areas in Figure 4.1 indicate the different interfaces between "reality" and "simulation".

4.2.1 Between Design/Layout as Process Simulation Input

According to the work flow outlined in Chapter 3, the layout of the masks is one of the two main inputs for process simulation.

Normally this layout is available in GDSII-binary format which can be read by any of the above mentioned commercially available TCAD tools. To mirror the activities carried out in the mask shop this data must undergo the same transformations as in reality listed in Subject 4 of

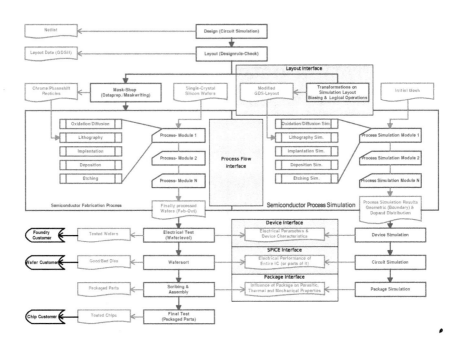

Figure 4.1: Scheme of identified interfaces between TCAD and semiconductor fabrication

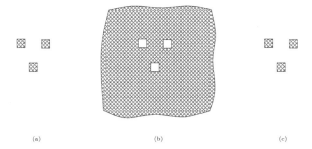

(a)   (b)   (c)

Figure 4.2: Example for a contact layer GDSII mask layer information
(which is of dark field type) (a), the corresponding chrome mask
recticle and resist shape (neglecting proximity correction) (b)
and the resulting etched contact holes in the oxide layer (c)

Section 2.1. The "Mask Generation Instructions" define special boolean operations to modify the mask data in a way that certain effects of wafer processing are cancelled out. Examples for possible corrections like simple mask biasing or proximity corrections are given in Section 3.3.2. An example for typical "Mask Generation Instructions" is given in Table 4.1. The mask levels are defined by the GDSII mask numbers given in the layout data file. The mask description identifies every generated reticle by name and is the reference for the lithography step in fabrication. The working plate field gives the data type for each mask. "DK" means *dark field* and "LT" means *light field*. The type gives an indication if the layout data has to be inverted for mask generation. Light field masks are masks which are generated as defined by the GDSII data. Dark field masks are generated with the *inverted* GDSII data. The contact mask is an example for a dark field mask. Since the contact mask has to be used for etching a hole (which is later filled with metal) into the oxide layer covering the silicon wafer, the photoresist covers all areas which are **NOT** contact. Therefore the GDSII data showing the contacts in the layout has to be inverted before mask generation. Figure 4.2 shows this situation. The *biasing* colummn give the bias to be applied to the GDSII data to get the final chrome mask dimensions. The strongest bias has to be applied to the mask level 10 listed in Table 4.1 which compensates for the effect shown in Figure 3.4 in Section 3.3.2. Since modern lithography steppers use 5x projection, the structures on the recticle are five times bigger than they are projected onto the photoresist. For mask quality control the critical dimensions of the chrome masks are defined in the "reticle CD" column in Table 4.1.

Figure 4.3 shows the detailed structure of the identified interface as motivated by the real mask generation and lithography process. The GDSII data is converted into the ASCII formatted CIF format (for easier processing) This data is then subject to boolean and biasing operations as defined by the mask generation instructions. To emulate the real shape of the photoresist a proximity correction is applied and the resulting contours are written back to a CIF format serving as mask information input during the process simulation.

A detailed description of the algorithm is given in Chapter 5.

RESIST TYPE: POSITIVE				
Mask Level	Mask Description	Working Plate Field	Biasing μm/side (1x)	5x Reticle CD (μm)
5	n-Well Mask	DK	0	3
8	p-Well Mask	LT	0	
10	Active Area Mask	LT	+0.08	2.4
8	p-Well Mask	LT	0	
24	VTP-Implant Mask	DK	0	
48	Anti Punch Through Mask	DK	0	
14	Gate Oxide	LT	0	
20	Poly 1 Mask	LT	+0.02	2
29	High Resistive Mask	LT	0	
30	Poly 2 Mask	LT	-0.04	3.0
21	n-LDD Mask	DK	0	
24	p-LDD Mask	DK	0	
23	N+ S/D Mask	DK	0	
24	P+ S/D Mask	DK	0	
34	Contact Mask	DK	+0.02	2.2
35	Metal1 Mask	LT	0	2.5
36	Via 1 Mask	DK	0	2.5
37	Metal 2 Mask	LT	0	3
38	Via 2 Mask	DK	0	2.5
39	Metal 3 Mask	LT	0	3
41	Via 3 Mask	DK	0	2.5
42	Top Metal Mask	LT	0	3
40	Pad Mask	DK	0	

Table 4.1: **Example of mask generation instruction table including biasing and CD (critical dimension) information**

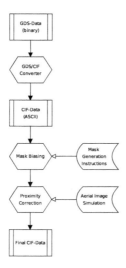

Figure 4.3: Interface of layout-data to simulation mask-data

4.2.2 Between Process Flow Description and Process Simulation Command File

The overall process flow information is typically documented in the database of a manufacturing execution system like PROMIS [149] from Brooks-PRI, SiView [150] from IBM, or WorkStream [151] from AppliedMaterials. The information relevant for process simulation inside this database is numerous, depending on the detailed level of the simulation models (as outlined in Table 3.1). Normally the following datasets are needed for process simulation:

1. Sequence of process steps representing the semiconductor process flow (oxidation $\Rightarrow$ layer thickness measurement $\Rightarrow$ implantation $\Rightarrow$ diffusion $\Rightarrow$ material deposition $\Rightarrow$ lithography etc.)

2. Blocks of process sequences which are carried out on the same semiconductor fabrication equipment and are normally organized as program sequences like oxidation/diffusion programs which can consist of up to dozens of single process steps with different temperatures (temperature ramps) and gas ambients (gas steps or ramps).

3. The detailed process parameter set of one single process step (e.g. ion species and composition, ion dose and energy, angle of incidence and rotational orientation of ion beam with respect to wafer, ion beam divergence, etc., for ion implantation)

4. Positions in the full semiconductor process flow where selected physical characteristics like layer thickness or sheet resistances are measured by using metrology tools on wafer level.

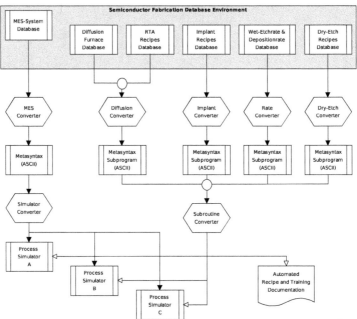

Figure 4.4: Interface of process flow information to process simulation command files

Subject 1 through Subject 3 represent the hierarchy levels from highest to lowest. It is advisable to use an abstract representation of data as an intermediate format between the process step information and the simulation command file. Since commercial simulators and simulators from university are still under heavy development, syntax changes of the simulator command files are happening frequently. Therefore a direct translator between process flow information and simulation is inflexible and difficult to extend. With an intermediate abstract format different types of simulators can be supported by a single source of data (see Figure 4.4).

Typically the Manufacturing Execution System (MES) is not storing the full details of the diffusion and oxidation recipes, the plasma etch programs or the etch and deposition rates of the wet chemistry used during semiconductor processing. This information is stored in separate databases as outlined in Figure 4.4. Normally there is only a reference to a machine recipe or a etch sink given in the MES flow. Again the strategy of exactly mirroring the real situation was chosen to set up the interfaces. The recipes are transferred by a converter into the meta-syntax and are then converted into the simulator syntax of the process simulator chosen. This procedure is shown on the left hand side of Figure 4.4. The converted recipes are then transferred into a subroutine format, provided by every commercial or university simulator available. Thus this approach can be used for every TCAD environment available. Furthermore, through the use of a meta-syntax the interpretation of the data formats inside the manufacturing system (MES and recipe databases) and the conversion into the different simulator syntaxes can be treated in a more systematic way, since both tasks cannot interfere in a single program but are performed in a modular way. Last but not least the meta-syntax enables a very compact and concise overview about the details of the semiconductor process flow. Thus, this syntax can be used as a source for a very sophisticated process flow description for documentation and training (see Section 6.2). The details of the implementation and examples for the meta-syntax used are given in Chapter 5.

4.2.3 Between Electrical Test and Device Simulation

After finished processing of the silicon wafers the first electrical test is the measurement of simple test structures and devices organized in Process Control Monitors (PCM) in the scribe-lines of the wafer. These measurements are carried out on automated tester systems on wafer level. The measurement procedures are again hierarchically oriented in the following way:

1. Measurement program set up for actual technology node.

2. Subprogram defined for actual PCM (normally several PCMs are inserted in the scribe-lines).

3. Module for Device Under Test (DUT) consisting of single program statements measuring relevant electrical parameters.

4. Single measurement algorithms for e.g. CMOS threshold voltage, or diffusion sheet-resistance.

5. Single steps of carrying out the measurement algorithm for e.g. CMOS threshold voltage in saturation. These steps define how the device terminals are connected to the voltage and current sources of the automated tester and how the currents and voltages of the DUT are measured.

Subject 3 to Subject 5 are mirrored on the device simulation side to provide comparable electrical data of measurements and simulation.

Since the algorithms under Subject 4 and Subject 5 are not changing frequently (the algorithms under Subject 5 are fixed[1] with the hardware of the automated tester used and measurement algorithms defined under Subject 4 are only changing, if a completely new device type is introduced) these algorithms are not converted on a daily basis. The structure of the interface is shown in Figure 4.5. The subprogram conversion of the DUT modules is carried out much more frequently on a daily basis.

There are two main application areas existing for converting electrical test programs. First, this conversion is used for the automated generation of big device test-chips during process development including new device architectures. Second, the standard PCM structure measurement algorithms have to be converted to match the simulation results with the PCM measurements. Both methods are described in Chapter 5 in detail.

---

[1] these algorithms are typically provided by the tester vendor in the form of test libraries

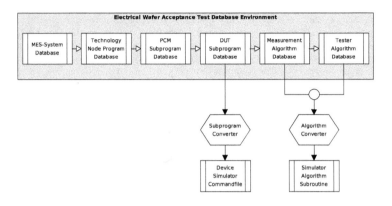

Figure 4.5: Interface of electrical wafer acceptance test information to device simulation command files

4.2.4 Between Device Characterization and Device Modeling (SPICE)

This interface deals with the generation of reliable device models for circuit modeling (e.g. SPICE). The main devices (NMOS/PMOS transistors for standard CMOS processes and, in addition, bipolar transistors for BiCMOS processes) of any new process fabrication must be characterized completely in terms of output characteristics, transfer characteristics, amplification, etc. This task results in scalable electrical models (BSIM3 for CMOS, VBIC for Bipolar transistors) or compact models for circuit simulation. In the TCAD fabrication integration scheme the source for this fitting procedure can be twofold: First, the usual way of measuring the characteristics on semiconductor wafer material and second, by simulating these characteristics with device simulation. The second approach has the enormous advantage of getting worst case predictions [152], [153], which are directly related to process parameter changes by applying statistical variations on selected semiconductor process step parameters (e.g. selected implantation doses). Furthermore, combined process and device simulations without the existence of any semiconductor material can generate preliminary models very early in the process development stage. Unfortunately the generation of SPICE models (e.g. BSIM3.3 or BSIM4 with hundreds of free parameters) is not an automated straightforward task. The process characterization engineers have to set up many initial values for starting the optimization of actual SPICE models and have to follow a complicated iterative strategy to get a good model with reasonable accuracy. Therefore an automated global optimizer for generating a good SPICE model is not available. Currently the only way to get so called "TCAD based models" is to generate characteristics with device simulation as they were measured on a real device and submit this information to process characterization engineers for the generation of SPICE models. Nevertheless, this approach enables the generation of a design environment of a new technology in a very early stage of a process development. The time to market for new process technologies is thus significantly reduced.

4.3 Package Modeling Interface

Since package modeling does not have a key focus in this work, no conversion tool was developed. However it would be beneficial to implement such a conversion into the overall TCAD flow. Since commercial package simulators provide a compact model (sub-circuit) of the parasitic elements introduced by the package, especially for RF and power applications this additional input could be very helpful. Currently this conversion is performed by hand or, more typically, not considered at all in the design process.

Chapter 5

Implementation

In the following the implementation of interfaces between fabrication and TCAD are described in some detail.

5.1   Layout Interface

The input data interface assumes two optional input formats of the layout data:

1. GDSII-Format: Binary format where the mask layers are identified only by an integer number between 1 and 64 (Advantage: Very compact / Disadvantage: Readable only by specialized layout viewers)

2. CIF-Format (Caltech Interchange Format): ASCII format identifying the mask layers by alphanumeric names (Advantage: Readable by any text editor / Disadvantage: Huge size for bigger layouts)

Since in semiconductor fabrication the mask layers are normally identified by alphanumeric names (like AA for active area mask) and the layout input for process simulations consists normally of only a very limited number of devices (typically one or two per simulation run) the CIF format was chosen as the standard input format due to the above mentioned advantages.

The GDS-layout input is initially converted to CIF-format by using a process node specific lookup-table which refers between the GDSII layer numbers and the mask level names.

In the Synopsys TCAD toolset the CIF-format is then transformed as described in Section 3.3.2 by setting up the mask transformations in a command file which is specific to the semiconductor process flow in scope. This setup file is only defined once in the development phase of the process and never change through the life cycle of the semiconductor process (otherwise every layout would have to be changed according to the modifications). An example for such mask generation instructions has been given in Table 4.1. The layout transformation is performed by simply starting the PROLYT layout tool in batch mode. Since normally not the whole layout of a device has to be simulated, only a selected portion of the device is defined in the layout and the according positional information and a name for this region definition is saved into the CIF-layout representation as additional layers recognized by the process flow simulator. The layer name identifies the dedicated simulation region. Therefore multiple simulation regions may be defined in a single layout. Finally the positions where measurements should be carried out during the process flow are marked as one-dimensional regions on the layout.

Optionally an aerial image simulation can be performed to generate mask information matching more closely the real shapes of the photo resist. Examples for such a preprocessing step can be found in Figure 3.5 in Section 3.3.2 and in Figure 6.6 and Figure 6.7 in Section 6.3. This preprocessing is mandatory, if capacitance coupling effects or crosstalk play an important role in the simulated structure. For instance the exact shape of the floating gate has to be resolved with high accuracy to calculate the exact coupling ratio of a tunneling EEPROM device as shown in Section 6.3.2.

This mask position and dimension information is passed automatically to the process simulator and defines the position and shape of the photo resist masks during the lithography simulation steps.

5.2 Process Flow Interface

There are levels of integration differentiated by the status of the semiconductor process flow.

5.2.1 Semiconductor Process Flows in Development

Typically these flows are subject to frequent changes during development. Therefore the detailed flow descriptions are not implemented in the MES system until the process flow is frozen. Normally the flow description is tabulated in a simple file format (like MS EXCEL or ASCII-table) for reference during the process development phase. This system supports also the fast setup of short process sequences by simply typing the relevant process steps into an EXCEL-sheet. An example screen-shot of this format is given in Figure 5.1.

By standardizing the flow description information for this stage, it is possible to define an automated converter which transfers the information identified in Section 4.2.2 into the abstracted semiconductor process simulator input language called SPR (Simple Process Representation) which is the input language for the commercial tool LIGAMENT. This tool generates the final process simulator input command files. This task is implemented via PERL-converter scripts. By using the SPR language it is possible to use various different process simulators even from different vendors like DIOS or SUPREM4.

5.2.2 Semiconductor Process Flows in Production

Once the semiconductor process flow is frozen and released, the MES system provides any input on the process flow which is necessary for process simulation. However this information must be reduced to the subset relevant for process simulation. This task is again performed with PERL-converter scripts and results in the table shown in Figure 5.1.

The overall scheme of conversion is outlined in Figure 5.2.

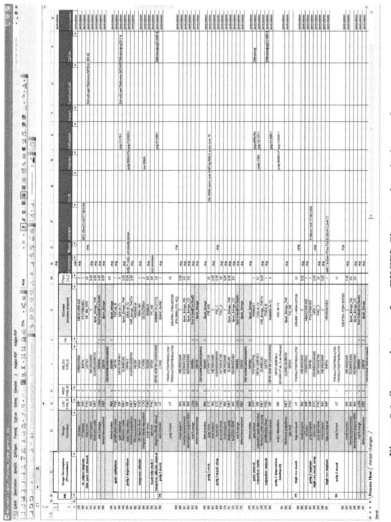

Figure 5.1: Screen-shot of EXCEL-Sheet description for research&development short-loops or semiconductor processes in development

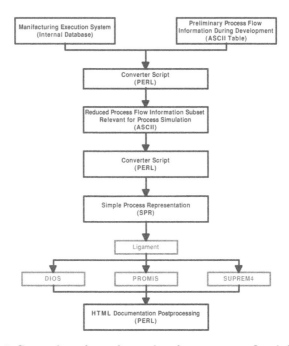

Figure 5.2: Conversion scheme for semiconductor process flow information

Two different data sources have been established for process flow conversion. First the MES database and second a simple ASCII table representation. Both sources are converted into a common reduced process flow information subset suitable to get a good, compact overview of the relevant process flow information. A second converter script is transferring this information into the simple process representation syntax as input to the TCAD framework tool LIGAMENT. This tool is able to generate different types of process simulator (e.g. DIOS or TSUPREM4) command files. After a full process flow simulation was performed the resulting structural and doping information can be used to generate a HTML documentation of every simulated process step.

5.2.3 Diffusion Recipes

One of the most important information to be provided to the process simulator in detail are the diffusion and oxidation recipes. These recipes are normally not stored in the MES but in a local diffusion furnace database. An example for such a system is the *Supervisor Station* from ASM. These systems have an internal format representation which has to be converted into a meta-syntax including the relevant information for the process simulator. An example of the original format of such a diffusion recipe is given in Listing 5.1. The listing shows the typical level of detail of a diffusion program. The program includes the analog output level and the alarm limits of the gas flow controllers, alarm patterns of the digital input channels and temperature and time settings of the diffusion program.

The compressed representation of the program shown in Listing 5.1 is obtained by the diffusion recipe converter as given in Listing 5.2 in the meta-syntax format. The first column gives the step number, the second the position of the quartz boat with the wafers. The third column shows the boat speed indicating how fast the boat is moved into and out of the quartz tube in the furnace. The next column shows the duration of the step. The following two columns give the target temperature and if the temperature has to be ramped with a certain ramp rate over the duration of the step. Finally the remaining rows give the targets for the gas flows set during the diffusion step.

The final simulator diffusion program is obtained by the meta-syntax to SPR converter as shown in Listing 5.3. This syntax is already very close to the process simulator syntax. However it does not include special model commands, specific to the process simulator. The syntax is targeted to give the generic inputs similar to every process simulator. Specific process simulator commands have to be defined with *include* statements.

The final simulator syntax is then generated by the TCAD environment.

5.2.4 Other Recipes

The conversion of the other recipes (implantation, etching and deposition) is quite straightforward. For implantation the database contains exactly the same information as needed by the process simulator. For isotropic etching and deposition only the etch rates and time are needed to simulate the change of geometry. The only not well developed interface is the anisotropic etch interface because of the unavailability of reliable and fast process simulators for this step. The equipment simulators needed for simulation of the effects of changes in the dry etch recipes on the overall etch profiles are still not available for routinely simulation tasks.

5.2.5 Conclusions

Because of the established system the turnaround time for setting up a process simulation flow has been reduced from days to hours. The probability of faults in the first setup of the process flow has been dramatically reduced. Especially the diffusion recipe conversion is inevitable to guarantee that a possible mismatch between simulation and measurements is not due to a wrong entered simulation recipe but because of poor calibration or new effects not covered by simulation.

```
 1 Step nr (00) Standby
 2    Message (1):STANDBY
 3    Analog Output (1):BubTemp to 20.0 degC
 4    Analog Output (2):N2-D to 10.00 Slm
 5    Analog Output (12):N2PrgFnc to 100.0 Slm
 6    Analog Output (14):TmpFlgLq to 85.0 degC        #TEMPFLANGE CONTROL
 7    Analog Output (16):TmpFlg to 85.0 degC          #READ ONLY
 8    Alarm Limit on Analog Output (1):BubTemp at 5% (+/- 5.0 $\deg C$  )
 9    Alarm Limit on Analog Output (4):O2-Low at 5% (+/- 0.075 Slm )
10    Alarm Limit on Analog Output (13):FRBlower at 0% (+/- 0.0 %  )
11    Alarm on Digital Inputs (1-8) 1110-011 (1):N2-Press (2):CA-Press (3):ExhGasc (7):
      ProcIntk (8):CabDoors
12    Alarm on Digital Inputs (9-16) 0101111- (10):ExhFurnC (12):OverHeat (13):H2O-reac
      (14):H2O-car (15):Blow-car
13    Alarm on Digital Inputs (17-24) 00000101 (22):FRH2OTmp (24):ExhLowCh
14    Alarm on Digital Inputs (25-32) -111---1 (26):FRAirTmp (27):Bub-Temp (28):Bub-Levl
      (32):HCl-Leak
15    Digital Outputs (1-8) 1XXXXX-X (1):N2-Orf
16    Temperature: Normal Table <100@950> entry 0 Profile table A
17 Step nr (01) Optie-Purge
18    Message (4):TIME DELAY      #OPTIONAL PURGE 3MIN
19    Analog Output (2):N2-D to 9.50 Slm
20    Analog Output (4):O2-Low to 0.500 Slm
21    Time: 00:00:10
22 Step nr (02) Boat-In
23    Message (2):BOAT IN/OUT
24    Boat to 1000 mm at 100 mm/min. Rotation 0 rpm
25    Wait for Boat to reach Setpoint
26    Wait for Digital Inputs (9-16) 0010000- (11):TubeClsD
27    Time: 00:12:00
28 Step nr (03) Stabilise
29    Message (3):STABILISE
30    Boat to 1000 mm at 100 mm/min. Rotation 1 rpm        #START ROTATION
31    Analog Output (11):N2LowCh to 5.00 Slm
32    Analog Output (12):N2PrgFnc to 25.0 Slm
33    Alarm on Digital Inputs (9-16) 0111111- (10):ExhFurnC (11):TubeClsD (12):OverHeat
      (13):H2O-reac (14):H2O-car (15):Blow-car
34    Alarm on Digital Inputs (17-24) 00000111 (22):FRH2OTmp (23):LowChPr (24):ExhLowCh
35    Abort on Any Alarm
36    Time: 00:15:00
37 Step nr (04) Heat-Up1
38    Message (11):HEAT UP      #to 900 C, 10 C/min
39    Alarm Limit on Analog Output (4):O2-Low at 0% (+/- 0.000 Slm )
40    Abort Recipe A1 <ABORTATM>
41    Abort on Digital Input Alarm (17-24) 00000011 (23):LowChPr (24):ExhLowCh
42    Temperature: Normal Table <100@950> entry 3 Profile table A
43    Time: 00:12:30
   ...
 99    Message (15):ANNEAL
100    Analog Output (3):O2-High to 0.00 Slm
101    Alarm Limit on Analog Output (3):O2-High at 0% (+/- 0.00 Slm )
102    Alarm on Digital Inputs (1-8) 1110-001 (1):N2-Press (2):CA-Press (3):ExhGasc (8):
       CabDoors
103    Time: 00:03:00
104 Step nr (14) Cool-Down1
105    Message (12):COOL DOWN      #to 900 C, 5 C/min
106    Boat to 1000 mm at 100 mm/min. Rotation 0 rpm
107    Analog Output (5):HCl to 0 Sccm
   ...
141    Message (2):BOAT IN/OUT
142    Boat to 10 mm at 100 mm/min. Rotation 0 rpm
143    Wait for Boat to reach Setpoint
144    Time: 00:11:00
145 End of Recipe
```

Listing 5.1: Original Diffusion Recipe Format

#Step number	Boatpos	Boatspeed	Time	Temperature	Temprate	H2	HCL	N2	O2
#	mm	mm/min	min	degC	degC/min	SLM	SLM	SLM	SLM
0	10	0	0.000	775.0	0	0	0	10	0
1	10	0	0.167	775.0	0	0	0	9.5	0.5
2	1000	100	12.000	775.0	0	0	0	9.5	0.5
3	1000	100	15.000	775.0	0	0	0	9.5	0.5
4	1000	0	12.500	900.0	10.00	0	0	9.5	0.5
5	1000	0	10.000	950.0	5.00	0	0	9.5	0.5
6	1000	0	0.167	950.0	0	0	0	9.5	0.5
7	1000	0	20.000	950.0	0	0	0	4.75	0.25
8	1000	0	0.500	950.0	0	0	0	4.75	5
9	1000	0	31.250	950.0	0	0	0.155	0	5
10	1000	0	0.500	950.0	0	0	0	0	5
11	1000	0	0.167	950.0	0	0	0	0	5
12	1000	0	0.500	950.0	0	0	0	18	5
13	1000	0	3.000	950.0	0	0	0	18	0
14	1000	100	10.000	900.0	-5	0	0	18	0
15	1000	100	12.500	775.0	-10	0	0	18	0
16	1000	0	0.167	775.0	0	0	0	18	0
17	990	20	1.000	775.0	0	0	0	18	0
18	970	5	12.000	775.0	0	0	0	18	0
19	10	100	11.000	775.0	0	0	0	18	0

Listing 5.2: Meta-syntax Diffusion Recipe Format

```
* procedure 11100

defop rec11100() {
    anneal(time : 12.500 min, temperature : {775.0,900.0} degC, pressure : 1 atm, nitrogen : 9.5 l/min
        , oxygen : 0.5 l/min)
    anneal(time : 10.000 min, temperature : {900.0,950.0} degC, pressure : 1 atm, nitrogen : 9.5 l/min
        , oxygen : 0.5 l/min)
    anneal(time : 0.167 min, temperature : 950.0 degC, pressure : 1 atm, nitrogen : 9.5 l/min, oxygen
        : 0.5 l/min)
    anneal(time : 20.000 min, temperature : 950.0 degC, pressure : 1 atm, nitrogen : 4.75 l/min, oxygen
        : 0.25 l/min)
    anneal(time : 0.500 min, temperature : 950.0 degC, pressure : 1 atm, nitrogen : 4.75 l/min, oxygen
        : 5 l/min)
    anneal(time : 31.250 min, temperature : 950.0 degC, pressure : 1 atm, hcl : 0.155 l/min, nitrogen
        : 0 l/min, oxygen : 5 l/min)
    anneal(time : 0.500 min, temperature : 950.0 degC, pressure : 1 atm, nitrogen : 0 l/min, oxygen
        : 5 l/min)
    anneal(time : 0.167 min, temperature : 950.0 degC, pressure : 1 atm, nitrogen : 0 l/min, oxygen
        : 5 l/min)
    anneal(time : 0.500 min, temperature : 950.0 degC, pressure : 1 atm, nitrogen : 18 l/min, oxygen
        : 5 l/min)
    anneal(time : 3.000 min, temperature : 950.0 degC, pressure : 1 atm, nitrogen : 18 l/min)
    anneal(time : 10.000 min, temperature : {950.0,900.0} degC, pressure : 1 atm, nitrogen : 18 l/min)
    anneal(time : 12.500 min, temperature : {900.0,775.0} degC, pressure : 1 atm, nitrogen : 18 l/min)
    anneal(time : 0.167 min, temperature : 775.0 degC, pressure : 1 atm, nitrogen : 18 l/min)
}
```

Listing 5.3: SPR Diffusion Recipe Format

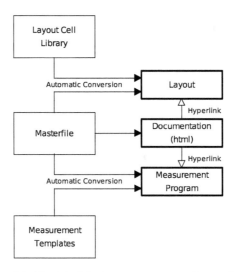

Figure 5.3: Electrical test program conversion work-flow

5.3 Electrical Test Interface

5.3.1 Introduction

A typical test-chip architecture demands numerous repetitions of the same test device with only slightly varied parameters. Traditionally the development engineer specifies the devices needed and documents their parameters. Based on this documentation the layouter draws the test structures and a test engineer creates test programs for an automated parameter tester. The large amount of test structures as, e.g., seen in a typical CMOS based high voltage process demands the need for an automated one-source template based system to create layout, documentation, and test programs. This system itself has to be flexible enough to allow easy incorporation of new test structures and measurement methods.

The development engineer creates one master file defining a set of test structures and parameter variations (such as e.g. well distance variations). In this file one can also define the measurements to be performed, using a simple meta-syntax. A script that uses a predefined cell library to create the layout of test lines with a standardized pad layout reads this file. The test lines are arranged and placed in the test chip automatically, and the exact position of these lines is provided to the automatic test system. A second script uses measurement templates to convert the master file in test programs suitable for the specific test system. The resulting layout files and test programs are hyper-linked into the documentation that can be used as a reference by the development engineer fur further analysis. An overview about this scheme is shown in Figure 5.3.

Parameter	Comment	Parameter	Comment
Gate	Gate pins	*Idcpl*	Compliance for drain current
Drain	Drain pins	*Vsubsforce*	Optional substrate bias
Source	Source pins	*Isubcpl*	Substrate current compliance
Substrate	Substrate pins	*Dtime*	Settling time before measurement is taken
Type	Specifies NMOS or PMOS	*Strategy*	Measurement strategy. Supported strategies are measure, integrate and average
Idstart	Minimum drain current to start drain voltage sweep	*Stepno*	For strategy average: number of points to average
Vdsstep	Step width of drain voltage sweep	*Steptime*	For strategy average: time interval between measurements
Vdspoints	No. of points for drain voltage sweep	*Vtsmeas*	Return value: measured threshold voltage
Vdscpl	Compliance for drain voltage		

Table 5.1: Parameters for threshold measurement routine

5.3.2 Electrical Parameter Test

The modules involved in creating the test programs for the electrical parameter extraction use several layers of abstraction.

5.3.2.1 Measurement Routines

On the lowest level there are the test routines for parameter extraction itself, e.g., a threshold voltage extraction or a breakdown voltage extraction routine. These routines contain all the instrument specific commands and calculations that are necessary to obtain device parameters. The routines can make use of a fairly large number of parameters specifying everything up to delay times, integration times, and compliance values (see Table 5.1).

However, in most cases the test program designer does not want to care about all these details, but wants to use a simple routine where one specifies the pads to use and some basic parameters specific to the DUT. Therefore a set of templates was introduced to provide a translation from an easy-to-use syntax to the more powerful but also more complicated measurement routine syntax (see Listing 5.4). These templates also make it possible to have several measurement commands call the same routine with different parameters. Thereby one may specify measurements with different current compliance values or settling times for different types of transistors.

The development engineer does not need to deal with the specific parameter set for every basic routine. If the actual templates are not sufficient for the task needed, the specialist responsible for electrical test can provide a new template.

```
<Name>
DRT_SAMPLE
</Name>
<Parameter>
Voltage=-5.5 V
BV_Current=1e-6 A
ILEAK_MAX = 1e-6 A
BV_MIN = 5.0 V
</Parameter>
<Common_pins>
P_Bulk = 16
P_Low=1
</Common_pins>
<PIN_TABLE>
Spacing P_High
0.1     2
0.2     3
0.4     4
0.8     5
</PIN_TABLE>
<Measure>
S = Spacing
ILEAK = ILEAK(P_High, P_Low, Voltage)
BV    = BVOX(P_High, P_Low, BV_Current)
</Measure>
<Calculate>
</Calculate>
<Design_Rule>
S_1 = Spacing@(ILEAK>ILEAK_MAX)
S_2 = Spacing@(BV<BV_MIN)
DR = min(S_1, S_2)
</Design_Rule>
<Result>
ILEAK[pA]
BV[V]
S[um]
DR[um]
</Result>
```

Listing 5.4: Template definition

```
. . .
<VTSLIN>
RESULT = VTSLIN(D, S, G, B, TYP)
VTSLIN("G", "D", "S", "B", 'TYP', 1e−6, 0.1, 5, 5.0, 0.1, 0.0, 1e
    −6, 0.01, 'M', 3, 0.01, 3, RESULT )
</VTSLIN>
<VTSLIN_Var>
</VTSLIN_Var>
. . .
```

Listing 5.5: Electrical test-program example

5.3.2.2 Test Program Definition

A test program in the development engineers' view consists of the following parts (see Listing 5.5).

- A header specifying the name of the test-line and thus the position of the line in the floor-plan

- Some global parameters for the whole test-line

- A list of devices specifying the associated pins and specific device information, such as geometric parameters

- A set of measurement instructions to be performed on each device in the line

- An optional set of instructions to be performed after the measurements. This block is intended for doing calculations on the measured parameters, such as extracting design rule parameters

This configuration offers the flexibility to define programs for design rule test-chips and similar test-chips consisting of test-lines with only similar devices. However, the test program has to be defined manually for each test line.

5.3.2.3 Automated Creation of Test Program Definitions

For a recent high voltage process development project the evaluation of geometric variations of high voltage transistors was needed. Because of the fact that these transistors have a rather complex layout and therefore a lot of layout parameters that modify the transistor behavior, the testchip turned out to include up to 1536 similar devices to be layouted in up to 129 test lines. In total there have been 466 test lines with only 7 different types of transistors. In order to rule out human errors and make life easier for designers and test program developers, it was crucial to automate as much work as possible.

As input only some basic device layout and tables were needed, to specify which parameters to vary in what sequence. From these input files the layout and arrangement of the devices on the test lines was performed automatically. The resulting arrangement list was used as an input for a script that used a general template measurement program similar to Listing 5.4 and filled up the <NAME> and <PIN_TABLE> section with the correct information. This configuration only requires the seven template files to be created and guarantees that all test lines are measured in the same way.

The resulting program definition files are converted into test programs for the test system. In addition the layout scripts create a coordinate list for the test lines, so that these coordinates can be appended to the wafer definition required by the test system.

5.3.3 Documentation

The whole test-chip is documented in HTML format by a table of test lines, each of which has a link to the layout and to the test program definition file. This documentation is also created from the arrangement list resulting from the transistor variation, and from a manually created arrangement list with those design rule structures that are only available on single test lines. However, the conversion to HTML is performed with a script and therefore is not prone to human errors.

5.3.4 Conclusions

The creation and the maintenance of test programs for design rule and device evaluation test-chips has been drastically facilitated by the introduction of an automated test program creation suite. The time to create a test program for device screening could be reduced from several weeks when the programs would have been created manually to a few days for the initial working templates. Any changes in the measurement sequence could be done within hours instead of weeks. It was guaranteed also, that all devices were measured in the same way without the risk of typing errors in the measurement programs.

5.4 SPICE Modeling Interface

The SPICE modeling interface is implemented (as outlined in Chapter 4 in Section 4.2.4) analogously to the typical approach how SPICE models are generated. Simulated characteristics of a high-voltage PMOS device used as a driver inside a EEPROM process technology together with the fitted SPICE model are shown in Figure 5.4. The most important characteristics necessary for generating a SPICE model (transfer characteristics and output characteristics as a function of the terminal voltages at gate, drain and substrate) are shown. The crosses denote the simulated current values and the solid lines denote the fitted SPICE model. From this graph it can be clearly seen that simulation results obtained from device simulations can be used as if they were measured results. The SPICE models obtained by this method have been as stable in circuit simulation as the measurement based ones.

With this method the circuit designers are able to start the design of an integrated circuit based on a process technology in development much earlier than with the conventional approach. The additional effort of generating SPICE models twice (first the TCAD based models and second

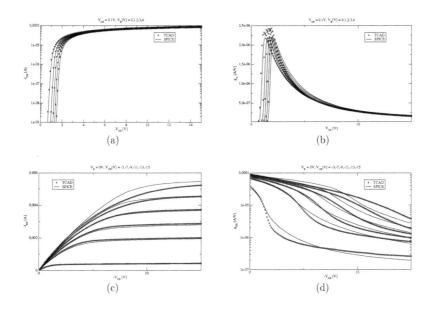

Figure 5.4: Transfer (a), output (b), transconductance (c) and conductance (d) characteristics of a high voltage PMOS driver transistor at different substrate and gate voltages respectively. TCAD simulations are compared to the results of the fitted SPICE model

the measurement based models when silicon material is available) is justified by the time gained during the design of the first product. Having a integrated circuit prototype ready early in the semiconductor process ramp-up phase is inevitable to introduce and stabilize new process technologies into the semiconductor manufacturing line. Furthermore, the manufacturability of a new process technology can be evaluated and improved already during the early development phase. Integration problems can be screened out much earlier thus reducing the development risk significantly.

Chapter 6

Industrial Application of TCAD

6.1 Transfer of Semiconductor Diffusion and Oxidation Process Recipes Between 4" and 8" Wafer Fabrication Facilities

TCAD has found to be very useful in reducing the risks of semiconductor process flow transfer between different fabrication facilities [154]. When transferring diffusion or oxidation process recipes from one type of equipment (e.g. 4" diffusion furnace) to another type of equipment (e.g. 8" diffusion furnace), it is generally not possible to copy the diffusion recipe without modifications. Especially, the temperature ramp rates for 8" diffusion recipes are usually significantly slower than for 4" equipment. The main reason for this difference is the different mechanical stability of 8" and 4" wafers. A plastic deformation of the 8" wafers called "Furnace Slip" is occurring during high temperature processing, if the temperature ramp rates are too steep [155]. A table of maximum allowable temperature ramp rates for a vertical 8" furnace is given in Table 6.1. Although, because of this constraint, the recipes might differ significantly, the impact on the wafer has to be nearly identical for 4" and 8" equipment. Thus optimization of the diffusion recipes is needed in order to make the differences in dopant distribution and oxide thickness between 4" and 8" recipes as small as possible.

The following procedure was followed to optimize the 8" recipes:

1. The original 4" recipe is changed according to the new maximum ramp rates allowed.

Temperature Range	Ramp-Up Rate	Temperature Range	Ramp-Down Rate
$750\,^{\circ}C$ - $1000\,^{\circ}C$	$7\,^{\circ}C/\mathrm{min}$	$1200\,^{\circ}C$ - $1150\,^{\circ}C$	$1\,^{\circ}C/\mathrm{min}$
$1000\,^{\circ}C$ - $1100\,^{\circ}C$	$3\,^{\circ}C/\mathrm{min}$	$1150\,^{\circ}C$ - $1100\,^{\circ}C$	$2\,^{\circ}C/\mathrm{min}$
$1100\,^{\circ}C$ - $1150\,^{\circ}C$	$2\,^{\circ}C/\mathrm{min}$	$1100\,^{\circ}C$ - $1000\,^{\circ}C$	$3\,^{\circ}C/\mathrm{min}$
$1150\,^{\circ}C$ - $1200\,^{\circ}C$	$1\,^{\circ}C/\mathrm{min}$	$1000\,^{\circ}C$ - $750\,^{\circ}C$	$3\,^{\circ}C/\mathrm{min}$

Table 6.1: Maximum allowable temperature ramp rates for vertical 8" furnaces

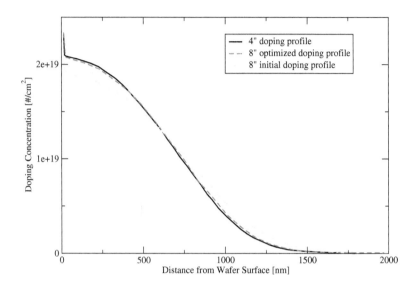

Figure 6.1: Doping profile for a 4" diffusion furnace compared to a 8" diffusion furnace before and after optimization

2. The main step contributing mainly to the overall thermal budget is identified.

3. A score function giving a minimization target for the optimization was defined.

4. The length of the main program step in the process simulation was varied in an optimization loop until the score function was minimized.

This algorithm was described already in Section 3.4, Figure 3.7. For optimization the framework SIESTA [144] was used. As a score function

$$\int_0^{x_{max}} |\ln N_{D,4''}(x) - \ln N_{D,8''}(x)| dx \qquad (6.1)$$

was chosen. x is the depth measured from the surface into the wafer, x_{max} is the maximum depth of the process simulation region. $N_{D,4''}(x)$ is the resulting doping profile from the process simulation of the 4" recipe. $N_{D,8''}(x)$ is the resulting doping profile from the process simulation of the 8" recipe.

In Figure 6.1 the initial and final 8" doping profile of a typical well diffusion recipe are shown. The resulting 8" diffusion recipe is shown in Figure 6.2 in comparison to the 4" recipe. The reduced temperature ramp rates of the 8" recipe can be seen clearly. For optimization of doping profiles with junctions (e.g. to the substrate) a different score function, the well depth $x_j = x(N_d = N_a)$ may be used. As an example a typical n-well diffusion program in a p-type substrate wafer is

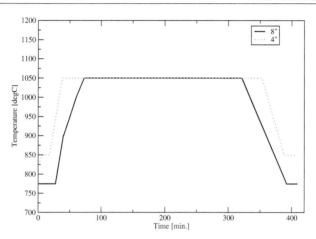

Figure 6.2: Graphical comparison between the 4" and 8" diffusion recipe for a typical p-well diffusion

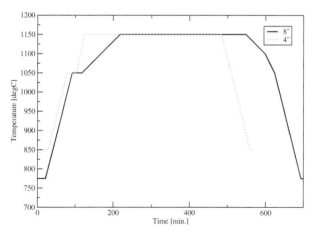

Figure 6.3: Graphical comparison between the 4" and 8" diffusion recipe for a typical n-well diffusion

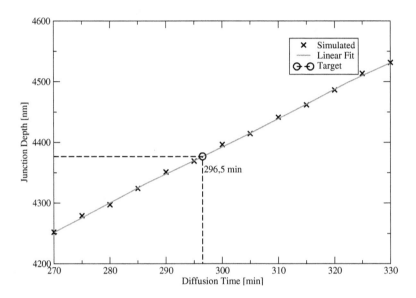

Figure 6.4: N-well junction depth over time of annealing step

shown in Figure 6.3 The resulting junction depths versus diffusion time are shown in Figure 6.4. Since the dependence of the junction depth on the diffusion time was exactly linear in this case, by fitting a linear equation through the simulated points and extracting the resulting annealing time, the 8" program can be optimized without any optimization loop like in the previous example.

6.2 Documentation and Training

The simulation results of the semiconductor process flow contain the information of the geometry of the simulated device (e.g., CMOS-Transistor) and the complete dopant distribution. This information is usually needed by process engineers but is very difficult, costly, or even impossible to obtain by other methods (e.g., SIMS, TEM). Combining the information of the process simulation and the process flow description results in a documentation of the process of very high quality. This information obtained by process simulation, which is usually only available for TCAD-Engineers, can be easily shared with process engineers, if a format with cross platform compatibility is used. One data format which fulfills the necessity of cross platform compatibility is the Hypertext Markup Language (HTML). Furthermore, especially during the development phase of a new process it is necessary that the simulation results can be transformed very quickly into HTML. A swift transformation of the process simulation results (which are usually in a platform dependent format or only viewable by special TCAD software) to HTML is achieved by a PERL script which extracts the relevant simulation results and links them with the description of the respective process step. The information of a diffusion or oxidation process recipe often consists of several dozens single process steps. This huge amount of information is best analyzed graphically as process temperature and gas flows versus time. Another information often needed by process engineers is the temperature versus time of the complete process which allows e.g. to figure out quickly the most relevant thermal process steps. Figure 6.5(a) shows a typical example for the documentation of a single diffusion program. Figure 6.5(b) demonstrates the documentation of a dedicated step during the processing of a bipolar transistor in a BiCMOS process technology. It shows the cross-section of the transistor after contact mask etch. Figure 6.5(c) indicates the total thermal profile of a process flow. The falling red line symbolizes the temperature of every diffusion recipe during the whole process. The time is the accumulated diffusion time of the diffusion programs. Figure 6.5(d) shows the one-dimensional cross-section through the source contact of a CMOS transistor.

6.3 Layout and Mask Generation

Based on the principles shown in Section 2.4 some examples for proximity correction of masks and possible applications are given in the following sections.

6.3.1 Proximity Correction of Masks

The proximity correction of masks may be of importance, if structure sizes approach the wavelength of the lithography system. At the 350nm node this is the case for the "Gate" mask, the "Active Area" mask, the "Contact" mask and the "Metal 1" mask. However in normal TCAD applications the proximity effects of these masks are not taken into account, because the typical critical dimensions of the front end masks ("Active Area" and "Gate" which are have the main impact on the device characteristics) are well controlled and are well calibrated in the TCAD simulations. The back end masks are only interesting for generating the contacts on top of the device. The detailed interconnect shape is not of interest for routinely TCAD simulations. However, in the application area of RF and high voltage, the exact interconnect shape is influencing the analysis strongly in certain aspects. The equivalent RLC network of the digital interconnect may impact the overall switching speed strongly (at least at ground rules below 180nm) [156],[157],[158]. For high voltage ultra low ohmic driver arrays with on-resistances in the milliohm range, the metalization resistance is contributing more than 50% to the total on-resistance.

These examples show, that an exact shape of the interconnect wires may impact the overall simulation result quite strongly. To obtain structures to analyze these influences more thoroughly with simulation, first a proximity corrected layout has to be generated. The detailed physics behind the generation of the corrected layout was described already in Chapter 2. For the following examples a modified version of LAYGRID [159],[160], the structure generator for

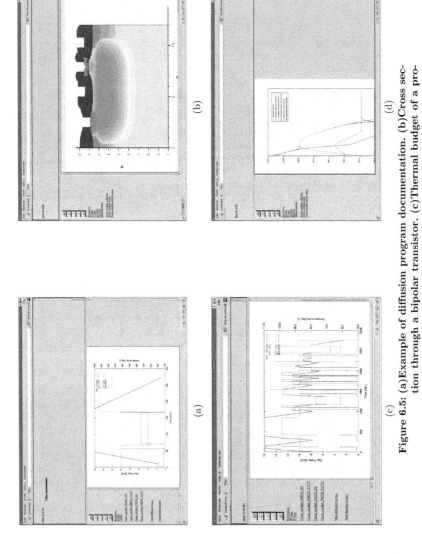

Figure 6.5: (a)Example of diffusion program documentation. (b)Cross section through a bipolar transistor. (c)Thermal budget of a process flow. (d)One dimensional doping cross section

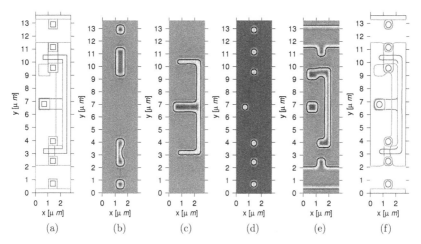

Figure 6.6: Initial layout of a digital inverter structure (a), contour plots of intensity distribution during the illumination of photo resist at different levels (b)-(e) and resulting resist contours after development (f)

the finite element electro-thermal simulation tool SAP [161],[162],[163] was used. The modification included the implementation of the aerial image simulator LISI developed by Heinrich Kirchauer [44] into the LAYGRID software. The original (mask biased) CIF file was taken together with the parameters of the lithography system (aperture etc.) as outlined in [164] and [165] and submitted to the modified LAYGRID code. The implemented LISI code generated a contour of every layer in the CIF file comprising of the light intensities at the surface of the photo resist (the aerial image). To obtain a fast and efficient simulation methodology the complicated and time consuming calculation of the exact photo resist shape after exposure and development was neglected. A certain threshold of the illumination intensity of the aerial image was chosen and the iso-contours of this intensity were extracted from the aerial image. This threshold was chosen to match the width of the final CDs of isolated mask lines accordingly. The contours were then written back into CIF file for further processing. The resulting CIF format can be used by any commercial or university TCAD simulator for further processing (e.g. process simulation). Examples for two digital cells (an inverter and a bigger digital cell comprising of 22 CMOS transistors) are shown in Figure 6.6 and Figure 6.7.

The resulting mask information was used to generate a three dimensional representation of the interconnect structures of the big digital cell with LAYGRID. A comparison of the metalization, and gate lines with and without proximity correction is given in Figure 6.8.

This structure can be used for further analysis of capacitance coupling, extraction of the RLC components of the interconnect or the overall metalization resistance.

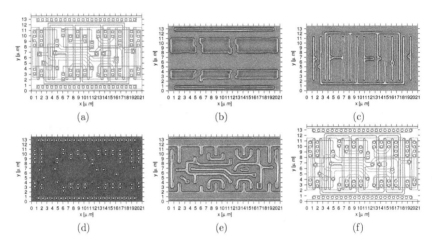

Figure 6.7: Initial layout of a big digital cell structure, contour plots of intensity distribution during the illumination of photo resist at different levels and resulting resist contours after development

6.3.2 Integrating an EEPROM Module into a State-of-the-Art Silicon Foundry Process with Three-Dimensional TCAD

6.3.2.1 Introduction

Non-volatile memories (NVM) [166],[167],[168] play an important role in modern System-on-a-Chip (SoC) solutions. The increasing demand of user-programmable information in such systems has led to new challenges in designing circuits with a certain amount of memory. NVMs are typically used in mobile, small systems for flexible applications which require variable information storage.

A variety of NVMs is available, each having different specifications according to the structure of the selected cell. A comprehensive overview is given in [169]. Two different programming principles can be identified, hot-electron injection (HEI) [170] and FOWLER-NORDHEIM (FN) tunneling [171],[172]. This work concentrates on an architecture that uses FN tunneling as the programming mechanism. This EEPROM cell was developed by J.M. Caywood [173],[174] and combines good endurance and reliability with a simple structure and good performance with average area consumption.

6.3.2.2 EEPROM Module Integration

The EEPROM p-channel memory cell was implemented in a common $0.35\mu m$ CMOS process flow. The front-end-process flow is presented in Figure 6.9. The detailed schematics for the

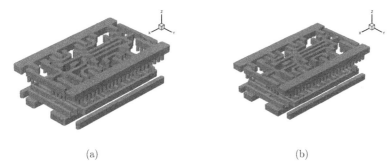

(a)   (b)

Figure 6.8: Comparison of the interconnect shape of a three-dimensional
structure of a big digital cell (a) with and (b) without proximity
correction

EEPROM process module steps may be found elsewhere [174]. Three "flavors" of the cell were evaluated. The implemented version makes use of a thick SiO_2 dielectric between the floating gate and the control gate (see Figure 6.10). Two other possibilities are a full ONO-stack as dielectric [175], or ONO between the control gate and the floating gate and SiO_2 for the control transistors [173].

Several implications arise for integrating an EEPROM memory in a CMOS process. First, the programming and erasing operation requires voltages up to 15V, which are normally far above the breakdown voltage of the S/D junctions (this is the case for technology nodes below $0.6\mu m$ and gets more severe for state-of-the-art nodes e.g. 130nm and beyond). Second, the added complexity of the overall process flow must not increase to a level where dual-chip packaging are cheaper solutions. As a consequence a maximum of only 2-4 additional mask alignments are acceptable. Third, the thermal budget of the high-voltage gate oxide for the control-gates will disturb sensitive threshold adjust implants and must therefore be placed before them. Fourth, for EEPROM memory operations additional high voltage devices are necessary to enable the generation of the programming voltage via charge pumps and to switch these voltages for the cell programming and erasing. As a consequence of these constraints the EEPROM-module must be integrated after the steps with the high thermal budget (e.g. the well diffusions) and before the sensitive threshold adjust and LDD steps which determine the standard CMOS logic. Since the base CMOS process offers already a dual gate (3.3V and 5V) analog mixed-signal option, the integrated flow includes three gate oxides. The HV-gate oxide of the cell is integrated right before the 3.3V and 5V gate oxides (refer to Figure 6.9). To get a deeper insight into the integration challenges, TCAD (Technology Computer Aided Design) simulations were used to find the best solution for the EEPROM module integration. Furthermore, the cell characteristics were optimized and the prediction of the electrical characteristics was used to generate preliminary SPICE models of the cell. This enabled a very early start of the memory block design. Additionally the transient behavior of the cell in programming operation was evaluated by TCAD.

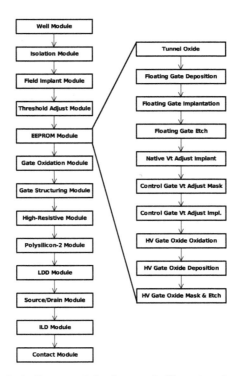

Figure 6.9: Block diagram of the front-end silicon foundry process flow

6.3.2.3 EEPROM Cell Optimization

To predict the EEPROM cell behavior two main areas of operation had to be investigated.
The accuracy of the DC characteristics of the cell is mainly determined by the overall calibration of the TCAD environment. Since this calibration was performed with the CMOS base process, the first results were already quite accurate.

The transient programming characteristics however, showed significant deviations from literature data [174]. The cause for these differences were inaccurate FN-tunneling model parameters in the device simulator DESSIS-ISE [83]. The most used model to describe tunneling is the FOWLER-NORDHEIM equation [176]

$$J = AE_{\mathrm{diel}}^2 \exp\left(-\frac{B}{E_{\mathrm{diel}}}\right) \tag{6.2}$$

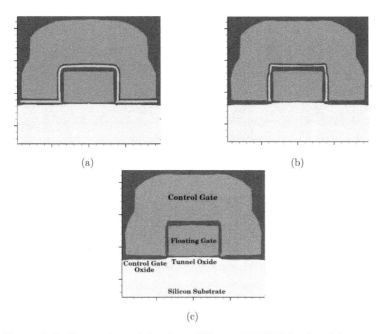

Figure 6.10: Comparison of the three different EEPROM cell architectures
(a) Full ONO Cell, (b) ONO-Spacer Cell, (c) Full Oxide Cell

which was originally intended to describe tunneling between metals under intense electric fields. The parameters A and B have been refined by LENZINGER and SNOW [177]:

$$J = \frac{q^3 m_{\mathrm{eff}}}{8\pi m_{\mathrm{diel}} h q \Phi_{\mathrm{B}}} E_{\mathrm{diel}}^2 \exp\left(-\frac{4\sqrt{2m_{\mathrm{diel}}(q\Phi_{\mathrm{B}})^3}}{3\hbar q E_{\mathrm{diel}}}\right). \tag{6.3}$$

This equation is the preferred model in the device simulator **DESSIS-ISE**. A comprehensive description of the tunneling mechanisms in semiconductors is given in [178]. The results of the calibrated model and the comparison to measurements are shown in Figure 6.11. The differences between the measurements and simulation are caused by the assumptions in (6.3), namely zero temperature, a triangular energy barrier, and equal materials on both sides of the dielectric. Nevertheless, this model is chosen on default to ensure stability and good convergence behavior in the TCAD device simulation.

The measurements were carried out on structured wafers with the tunnel oxide and a simple dot-masked polysilicon layer on top. The polysilicon dots were contacted with one needle of a micromanipulator, and the voltage between this contact and the wafer-chuck was varied appropriately.

One key parameter of an EEPROM cell using FN-tunneling is programming speed. Since the tunneling current density and therefore the programming time of the floating gate depends

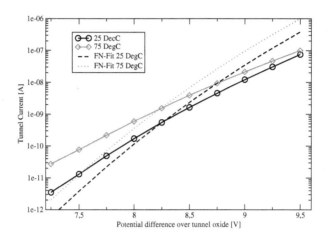

Figure 6.11: Comparison of simulated and measured tunnel currents through the tunneling oxide

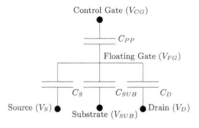

Figure 6.12: Capacitive equivalent circuit of the EEPROM cell

exponentially on the applied potential, c.f. (6.3) the gate coupling ratio[172]:

$$\alpha_g = \frac{\delta V_{FG}}{\delta V_{CG}} = \frac{C_{PP}}{C_{tot}} = \frac{C_{PP}}{C_S + C_{SUB} + C_D} \tag{6.4}$$

is the major contributor to cell speed. In order to optimize the cell speed, the Control-Gate/Floating Gate Capacitance C_{pp} must be maximized. Figure 6.12 shows all contributions to the coupling ratio.

The coupling ratio of the EEPROM cell is already excellent, since the special layout [174] enables an encapsulation of the floating-gate by the control-gate on all sides. The main parameter left for increasing the coupling-ratio is the thickness of the floating gate. However, there is a tradeoff

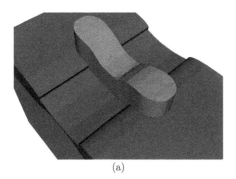
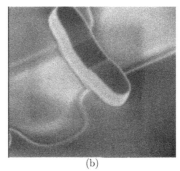

(a)   (b)

Figure 6.13: EEPROM cell after floating-gate mask etch (a)Three-dimensional **TCAD** process simulation (b) **SEM** photograph of real structure

between floating gate thickness, step-coverage, and minimum cell distance in a memory block. Using three-dimensional TCAD-process- and device-simulations a parameter optimum, matching the measurements, was found. Furthermore, the coupling-ratio of the cell itself could be predicted. The process simulation was again calibrated by comparing the three-dimensional TCAD boundary model with SEM pictures taken during fabrication (Figure 6.13).

The final structure of the EEPROM cell obtained by three-dimensional process simulation, which serves as input for a finite element analysis [162] for the extraction of the capacitances and the coupling-ratio, is shown in Figure 6.14.

The process simulation was performed by combining the simulation tools **DIOS-ISE** [82] and **TOPO3D** [179] [180]. The formation of the field oxide was carried out by a two-dimensional simulation performed with **DIOS-ISE**. Due to the three-dimensional nature of the problem, switching to a full three-dimensional analysis is required, beginning with the formation of the floating gate. Therefore the two-dimensional structure generated by **DIOS-ISE** was expanded to a three-dimensional geometry representation. In the following an isotropic deposition of the poly-silicon layer was performed with **TOPO3D** which is a rigorous three-dimensional simulator for etching and deposition processes. In order to transfer the floating gate mask, an etching model of **TOPO3D** was applied, which is capable of taking aerial image information into account. The aerial image figure of the floating gate mask was produced by the aerial image simulator **LISI** [181],[182],[183] and loaded into the topography simulator. Worth mentioning is that the mask information for the aerial image simulation is taken from a gds2-file containing a 3×3 cell array to prevent disturbances because of simulation domain boundaries (Figure 6.15). By this single process simulation step, five real process steps (mask deposition, mask illumination, mask development, anisotropic polysilicon etching and mask removal) are approximated in order to save computation time (without loosing significant accuracy as shown in Figure 6.13).
Subsequently (the doping formation steps are neglected, since they do not significantly impact the coupling ratio) the dielectric (SiO_2) between the floating gate and the control gate is grown. This oxidation process is approximated by an isotropic deposition step in the simulation analysis and is performed with **TOPO3D** as well, in order to avoid a full three-dimensional oxidation

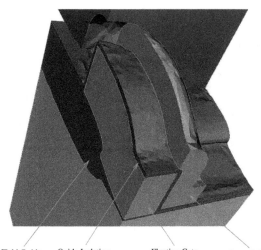

Field Oxide   Oxide Isolation   Floating Gate   Control Gate

Figure 6.14: Structure of the EEPROM memory cell generated by three-dimensional process simulation (one quarter of the cell is shown)

C_{PP}	Control G. $\leftrightarrow$ Floating G.	$2.38 \cdot 10^{-15}$F
C_{Sub}	Floating G. $\leftrightarrow$ Substrate	$5.69 \cdot 10^{-16}$F
$C_S + C_D$	Control G. $\leftrightarrow$ Substrate	$3.37 \cdot 10^{-16}$F

Table 6.2: Extracted capacitances within the EEPROM memory cell

simulation.

For the formation of the control gate equivalent simulation steps as for the formation of the floating gate were used. The simulation is finalized by deposition of a thick layer of silicon dioxide, which prevents against disturbances in the capacitance extraction due to boundary effects. The extracted capacitance values (listed in Table 6.2) give a gate coupling ratio of $\alpha_g = C_{PP}/(C_{Sub} + C_{PP} + C_S + C_D) = 72.4$ % (refer also to Figure 6.12 and (6.4)) which is in good agreement with measurements based on the algorithms given in [172].

6.3.2.4 Conclusions

TCAD methods are nowadays the method of choice for add-on module process integration. It was demonstrated that predictions for some technology key performance indicators can be derived. This methodology is excellently suited for a successful, timely and cost effective implementation of non-standard modules into a base process flow. In special cases three-dimensional process simulation is already feasible for industrial use.

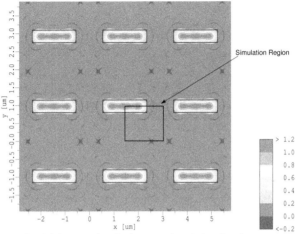

Figure 6.15: Aerial image simulation result of the floating gate mask of a 3×3 EEPROM cell array [181].

6.3.3   Generation of 3D-Mask Photo Resist Shape for lateral PIN-Diodes

This example deals with the coupled process and device simulation of a laterally diffused PIN-diode of special shape and subsequent comparison of the device simulation results to electrical measurements. Furthermore, it gives an outlook to layout optimization of laterally diffused devices in general. This example is one of the first fully integrated process and device simulations including non-Manhattan type structures and full incorporation of lithography proximity effects on photo resist level. Previous work was constrained to Manhattan type (90° angles between boundary primitives) structures without taking into account rigorous lithography simulation. By applying this new methodology significant differences in electrical characteristics between two-dimensional and three-dimensional simulations have been obtained. The simulated device was a Zener-diode in a $350nm$ CMOS technology process. This relatively "big" technology process was chosen to demonstrate the impact of three-dimensional effects in device characteristics even in such "old" process technologies. The layout of the element is of inherent two-dimensional nature because of the 45° angles in the p+ and n+ doped regions of the device. A complete process flow was simulated (see Figure 6.16) including the following critical steps:

1. Starting with two-dimensional process simulation up to the first critical two-dimensional lithography mask

2. Interfacing into three-dimensions by sweeping the structure laterally (see Figure 6.17)

3. Three-dimensional lithography simulation including simulation of resist exposure and development (see Figure 6.18)

4. Three-dimensional Monte-Carlo implantation (see Figure 6.19 (a))

5. Rapid-Thermal-Annealing

6. Mesh-refinement with respect to the metallurgical-junctions and the counter doped I-region of the PIN-diode

7. Placement of the metal contacts on top of the structure (see Figure 6.19 (b))

After Step 7 the simulation ends with a mesh representation of the final device including the three-dimensional doping distribution of the lateral PIN-diode (see Figure 6.19 (c)). To examine the difference between the two-dimensional simulation and the three-dimensional case, an additional two-dimensional process simulation was carried out up to contact deposition.

For the three-dimensional diode it turned out that phonon-assisted band-to-band tunneling cannot be neglected. In this steep pn-junction this effect is crucial to reproduce the measured reverse characteristics of this diode. The model had to be switched on, when the electric field in some regions of the device, exceeded approx. 8×10^5 V/cm. The band-to-band tunneling is modeled using the expression from SCHENK [184]. The results given in Figure 6.20 show clearly that this PIN-diode is a true three-dimensional device which cannot be simulated in two dimensions. The two-dimensional simulation predicts the breakdown voltage far too high (about 25 %) compared to measurements. The result of the three-dimensional analysis is in excellent agreement with the measured data. Figure 6.21 shows the resulting doping distribution after process simulation and its potential distribution after device simulation during onset of breakdown. In addition, due to the three-dimensional shape of the electric field inside the active region of the Zener-diode and the field peaks located at the border of the overlapping diffusions (see Figure 6.22), the tunnel current below breakdown is predicted about 4 decades to small compared to 3D results and the measurements.

6.4 Inverse Modeling of Polycrystalline Fuses

6.4.1 Introduction

For deep-sub-micron semiconductor process technology, the use of Polysilicon fuses, as one-time-programmable devices providing memories up to several kilobits offers a cheap, efficient, and area-saving alternative to small non-volatile memories for System-on-a-Chip solutions. Approaches to increase the memory density by using 3-state fuses of layered materials are also reported [185]. Another important application is the use in simple field programmable gate arrays or for trimming CMOS circuits for specific analog performance [186]. Furthermore, the fuses are used to provide variable elements as trimmable resistor or capacitor arrays [187]. Finally, the fuses may act as the classical protective elements for improved protection and replacement of critical components before actual failures [188]. Programming is performed by sending a broad current pulse through the fuse, resulting in an open-circuit after transition to a second-breakdown state. The transition occurs when parts of the Polysilicon layer reach the melting point, and the molten Silicon is transported from the negative end through drift of ions in the applied field [189]. Fuses implemented in deep sub-micron technologies become more and more attractive in terms of power and area consumption, and hybrid approaches using other materials are getting less important [190]. Nevertheless, going to smaller ground rules below 350 nm, implies decreasing supply voltages to 1.5 V and below [191]. This constraint requires a careful

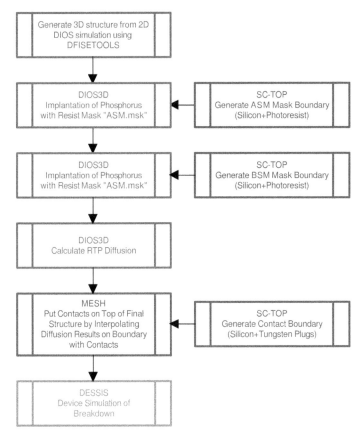

Figure 6.16: Schematic flow for coupled three-dimensional process and device simulation

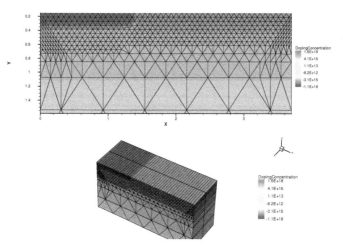

Figure 6.17: Two-dimensional initial structure and resulting three-dimensional mesh after conversion into three dimensions

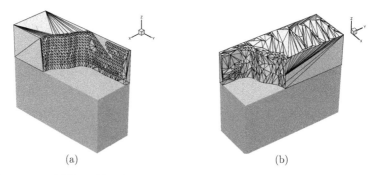

Figure 6.18: (a)SC-TOP simulation of ASM implantation mask shape (b)SC-TOP simulation of BSM implantation mask shape

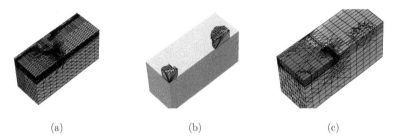

(a) (b) (c)

Figure 6.19: Generation of Zener diode mesh suitable for device simulation (a)Device after successive P+ and N+ Implantations using photo resist masks shown in Figure 6.18 (b)Boundary from interconnect simulation using SC-TOP for placement of the contacts (c)Final PIN-diode structure with merged contacts on top (ready for device simulation)

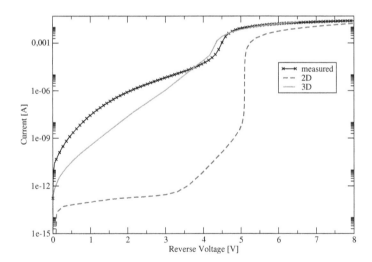

Figure 6.20: Comparison of two-dimensional and three-dimensional device simulation results with measured characteristics of PIN-diode

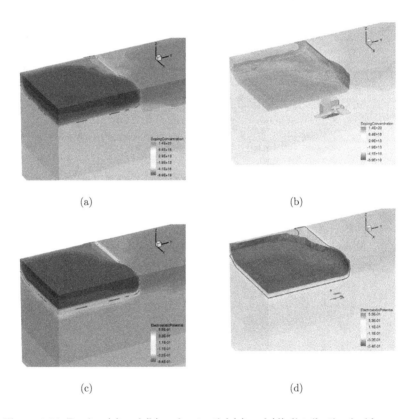

Figure 6.21: Doping (a) and (b) and potential (c) and (d) distribution inside the zener structure

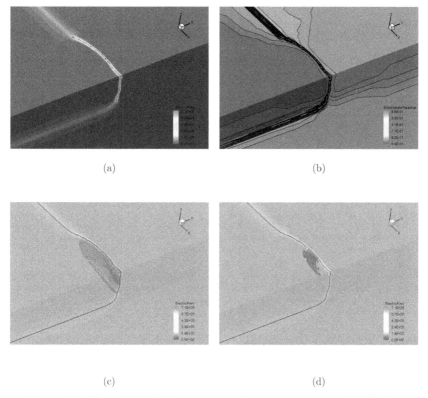

(a)   (b)

(c)   (d)

Figure 6.22: Electrical field distribution inside zener structure, (a) field distribution, (b) contour plots, (c) iso-surface at 0.25 MV/cm, (d)
iso-surface at 0.6 MV/cm

optimization of the fuse layout, ensuring an efficient and reliable programming mechanism [192] and minimizing the necessary power consumption of the fusing process. As the fusing process takes place in a short time interval (between a couple of nanoseconds up to the microsecond range), direct thermal measurements of this process are quite hard to obtain. Previous work [193] already shed some light on the physics behind the fusing mechanism, but the optimization of the fuse structure for reliable and fast fusing was only possible via expensive experimental work by using test chips.

This work focuses on gaining better insight into the materials characteristics used in the structure, to enable a layout optimization through simulation. Since the electrical and thermal properties of Polysilicon are a complex function of Polysilicon film doping, grain size, and grain morphology[194], the average electrical and thermal properties as a function of temperature were obtained by experimentally measuring the transient resistivity response of the fuse through Joule self-heating, and subsequent inverse modeling the measured data, to fit the observed behavior. The electro-thermal self-heating simulations were performed with the Smart-Analysis-Package (SAP) for three-dimensional interconnect simulation [195] in combination with SIESTA, a TCAD optimization framework combining gradient based and genetic optimizers[143]. This approach enabled the optimization of the fuse layout by significantly saving costs normally spent in design and production of layout test chips. Furthermore, a better insight into the transient electro-thermal effects occurring in the first couple of microseconds was gained.

6.4.2 Experiment

The fuse devices were fabricated in an industry-standard deep sub-micron polycided gate CMOS process. On a specialized test chip multiple layout variations were placed to find the optimum layout for fast and reliable fusing. A more complicated example of a fuse structure is shown in Figure 6.25. The first experiments were performed with rectangular pulses. Nevertheless, through the steep slope of the fuse terminal voltage the initial time regime of the fuse heating is not well resolved. Furthermore, the initial transient behavior of the measurement circuit yields high errors in the measured current data. To overcome these problems a voltage ramp was applied and the resulting fusing resistance was calculated by assuming ohmic behavior. The Polysilicon layer in the fuse is doped to solid solubility, and therefore its conductivity may considered to be approximately ohmic. Since all materials in the fuse except the Polysilicon layer are metallic, this assumption shall give a reasonable estimate for the fuse resistivity. The devices were stressed with different triangular voltage ramps for a few microseconds. A pulse generator was used to define the length of the pulse. As the generator has a typical output impedance of 50 Ω and the resistor of the Polysilicon fuse is lower than that, the source has to be buffered by an operational amplifier with a high slew rate to get a stable voltage. In order to avoid an additional voltage drop on a shunt resistor a current probe was used. In addition, the voltage on the fuse was monitored by an oscilloscope to calculate the right resistor value. The measurement principle can be seen in Figure 6.23.

The resulting measurement data for three different source voltages as a function of time are shown in Figure 6.24.

The resistance difference between the three voltages is caused by self heating of the whole structure (including the contact barrier layers) in the first microsecond of the applied pulse. The negative temperature coefficient of the resistance in all three curves occurs through the

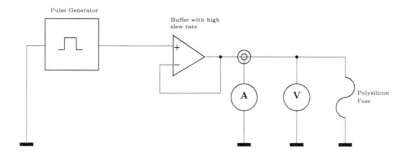

Figure 6.23: Schematic of the Polysilicon fuse measurement

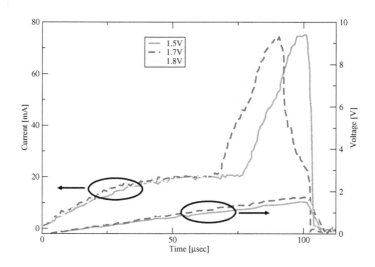

Figure 6.24: Measured current through fuse and voltage at the fuse terminals as a function of time

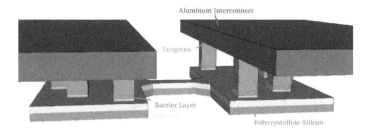

Figure 6.25: Fuse device structure showing the variety of included materials

combined Joule self-heating of the Polysilicon/Polycide layer sandwich (see Figure 6.25). The high noise in the data during the first $10\mu s$ originates from the low voltage level in this time regime and the resulting low signal-to-noise ratio.

6.4.3 Simulation and Inverse Modeling

6.4.3.1 Mathematical Models

For the numerical calculation of Joule self-heating effects two partial differential equations have to be solved. The continuity equation

$$\nabla \cdot (\gamma_E \, \nabla \varphi) = 0 \tag{6.5}$$

gives the electric potential φ where γ_E denotes the electric conductivity. The power loss density p is obtained by computing $p = \gamma_E(\nabla \varphi)^2$. The heat conduction equation

$$c_p \, \rho_m \frac{\partial T}{\partial t} - \nabla \cdot (\gamma_T \nabla T) = -p \tag{6.6}$$

is solved to obtain the temperature distribution where γ_T represents the thermal conductivity, c_p the specific heat, and ρ_m the mass density. The temperature dependence of the conductivities is modeled with

$$\gamma(T) = \frac{\gamma_0}{1 + \alpha(T - T_0) + \beta(T - T0)^2} \tag{6.7}$$

where γ_0 is the thermal or electrical conductivity at the temperature T_0, α and β are the linear and quadratic temperature coefficients of the specified materials [162].

6.4.3.2 Simulation Setup

The layout of the fuse was transformed into a three-dimensional representation of the device using a detailed process description of the interconnect forming deposition and etch steps. With the well known electrical conductivity of the interconnect and barrier layers the structural setup was checked by calculating the overall resistance of the structure excluding self-heating effects.

The Polysilicon conductivity was matched to the observed overall resistance and the resulting value was compared to independently measured sheet resistances of the polycrystalline layer in fabrication, resulting in an excellent agreement between the ohmic simulation and the measurements. The subsequent transient simulations were set up including the thermal coefficients of the electrical conductivity, the thermal conductivity, and the heat capacity of all layers in the structure. The starting values of these parameters were taken from literature data [196], [197], [198].

6.4.3.3 Inverse Modeling

The simulation framework SIESTA provides a wide range of optimizers that can be chosen to fit best for the current problems. Reference data for this optimization are measurements of the resistance calculated from Figure 6.24. At start time SIESTA provides the initial values of the free parameters for the three-dimensional interconnect simulator STAP of the SAP package, as introduced in [195]. The output of the simulation is parsed by SIESTA in order to compare it with the reference data. It produces a score value that indicates how good these two data sets match. This value is submitted to the optimizer which generates corresponding to the score value the next n-tuple of free parameters to improve the next score value that will be evaluated after the next simulation run with the currently produced values.

Wide interval ranges of free parameters can result in convergence problems because of non-physical parameter values which would cause negative resistance or negative doping. To avoid this, the simulation framework SIESTA provides a kind of divergence detection where SIESTA is signaled when the simulator has problems to converge. This feature allows the user to expand the intervals of the free parameters in a larger range as before.

6.4.4 Results and Discussion

The simulation framework SIESTA has to fit the thermal parameters of the electrical and thermal conductivities in order to minimize the difference between the reference and the simulation. To check the consistency of the setup, all thermal and electrical parameters were used for the automated simulation run, resulting in a total of 10 parameters. The best fit to the measured reference data is given in Table 6.3. The electrical and thermal conductivities $\gamma_{0,E}$ and $\gamma_{0,T}$ as well as the linear temperature coefficient of the thermal conductivity α_T for Polysilicon are in excellent agreement with data reported in [194]. The electrical conductivity of the Polysilicon/Tungsten Silicide sandwich as a function of temperature is comparable to data measured electrically by external heating of the layers.

The optimized parameter set results in resistance characteristics as shown in Figure 6.26 where an excellent match with the measurements is obtained. The strong increase of the resistance as a function of time because of self heating can be clearly seen. In addition, after a certain critical temperature is reached, the resistance drops dramatically and the ohmic approximation looses its validity. In order to generalize this result to other fuse geometries, this critical temperature has to be extracted. As expected, the critical temperatures of all the three samples are inside of a small interval about 1150 K (cf. Figure 6.28). This value is much smaller as the single crystal Silicon melting point of 1414 °C and the Tungsten Silicide (WSi$_2$ phase) melting point of 2015 °C[199]. The maximum temperature of the Polysilicon fuse is observed in the center of the Tungsten Silicide layer as shown in Figure 6.28.

	Poly Si	WSi$_2$
$\gamma_{0,\mathrm{E}}$ [1/$\mu\Omega$m]	0.12	1.25
α_{E} [1/K]	9.1×10^{-4}	8.9×10^{-4}
β_{E} [1/K^2]	7.9×10^{-7}	8.1×10^{-7}
$\gamma_{0,\mathrm{T}}$ [W/Km]	45.4	119.4
α_{T} [1/K]	2×10^{-2}	2.98×10^{-2}

Table 6.3: Parameters of electrical and thermal conductivity for materials used in the polysilicon fuse structure

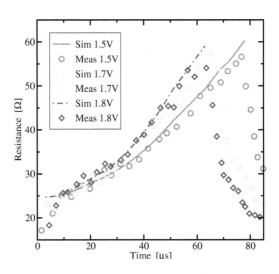

Figure 6.26: Comparison of measured and simulated polysilicon fuse resistance as a function of time

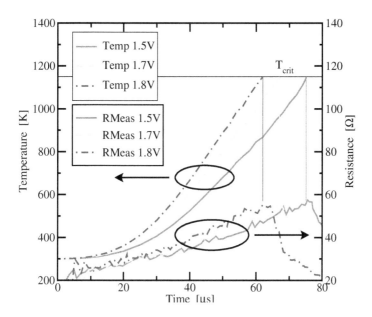

Figure 6.27: Comparison of the simulated temperature and measured polysilicon fuse resistance showing the extracted critical temperature

Several mechanisms for this low critical temperature are possible. First, the disordered region between the Tungsten Silicide and the Silicon may have a stoichiometry closer to the eutectic point of the Tungsten-Silicide system and therefore a lower melting point. But since the lowest eutectic temperature of the W-Si system is 1389 °C[199], this is not likely for pure alloys. Second, the high doping concentration of the Polysilicon layer reduces the melting temperature as reported for Silicon glasses with high Boron and Phosphorus contents. Finally, the assumption that all materials show Ohmic behavior over the full temperature range between 300 and 1200 K does not hold for higher temperatures.

The intended target to optimize fuse layouts for better performance is not affected, since it is obvious from Figure 6.27 that the melting begins always at approximately the same temperature. Therefore, the method should be applicable for other geometries as well. The agreement between experiment and simulation is excellent and provides a reliable base for carrying out predictive simulations of the transient temperature distribution during the initial heating phase of the fusing.

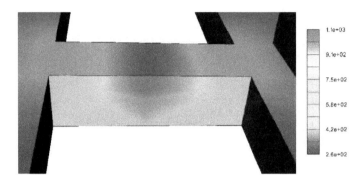

Figure 6.28: Temperature distribution in the polysilicon fuse interconnect structure at $65\,\mu s$ and $1.7\,V$

6.4.5 Conclusions

A method for obtaining important material parameters by inverse modeling using finite element simulations of complex interconnect structures was presented. This method is capable of describing the electrical behavior of interconnect materials over a significant temperature interval. Furthermore, it uses the transient thermal self-heating effect to separate different materials and their electrical and thermal properties. Nevertheless, the exact conduction mechanism inside the Polysilicon layer is still not well reflected in this analysis. In further work the impact of the grain boundary barriers and their behavior at high temperature has to be addressed by implementing a more accurate model like the model of Mandurah [200]. It was demonstrated that the method is consistent and gives an excellent match to experimental results. With the extracted critical temperature, where the material looses its ohmic properties, the geometry can be optimized in terms of reliability and speed.

6.5 Using TCAD with SPC

6.5.1 Introduction

A semiconductor fabrication process consists of several hundred unit process steps, each of which is subject to potential misprocessing. Such misprocessing typically occurs when wrong tool recipes are loaded and executed, or process steps are accidentally left out or performed twice. Although many of these issues are immediately detected at the next process step because of the physical deviation of the wafers from their usual appearance, some of these misprocessed wafers make their way through the whole production line and the failure is only detected at electrical parameter test. In such cases it is paramount that the cause of the failure is identified as quickly as possible to prevent other wafers in the fabrication line from being misprocessed the same way.

One special group of unit processes that is of particular interest in this context is the group of implant steps. Advanced semiconductor process flows contain several dozens of different implant steps, and since the implants do only affect the electrical, but not the mechanical or optical properties of the semiconductor wafers, a missing or double implant will typically be detected only at electrical test. Although large efforts have been made to prevent implant accidents, a certain risk remains in every not fully automated semiconductor fabrication facility.

Unfortunately, the relationship between the implants performed and the electrical behavior of the semiconductor devices is of high complexity, and the inference from the electrical data obtained at test to what actually happened during production requires the judgment and experience from device engineering experts.

In the following an approach is shown, how this inference can be made by a broader range of personnel with an even higher level of certainty.

During recent years, simulation techniques for semiconductor processing have been developed at a breathtaking speed. It is therefore feasible today, to feed a process flow, including all relevant process parameters, into a TCAD simulation, thereby creating a virtual semiconductor device such as a typical transistor, and extract the electrical properties of this simulated device. Parameters such as thresholds, saturation currents, sheet resistances or similar can thus be calculated for almost any given process flow.

So far, TCAD simulation has been extensively used for process development, but its application for manufacturing control and corrective action was very limited, which is partly due to the fact that its application needs skilled specialists. The advantages, of running a complete set of TCAD simulations of a transistor device for the process of record (POR), and for all process flows that result from both, an accidental missing and double implant (for each implant step), are described in the following sections.

6.5.2 Computational Effort

The current work is based on the analysis of austriamicrosystems' $0.35\mu m$ CMOS-mixed-signal process licensed from TSMC. This industry standard process offers two different gate oxide options (3.3V and 5V) resulting in 4 basic CMOS devices.

Performing a full factorial simulation of only 3 parameters (p-well, n-well and PMOS threshold adjust implant) with parameter values 0,1,2 (corresponding missing, correct and double implants) takes $3^3 = 27$ different simulation runs. Because of speed and memory constraints only one CMOS-transistor can be simulated per run, 108 different runs have to be executed to get the full information for these three implants on four transistors. This took a full weekend on a cluster of four 2GHz Linux computers, but having these data calculated up front, enables an engineer to identify very quickly the step where misprocessing occurred, in case a lot fails at electrical parameter test. Furthermore, this information has to be calculated only once, since these data reflect the situation in a frozen process flow.

6.5.3 Selecting a Set of Parameters

Since this $0.35\mu m$ process contains 16 implants in total, it is obvious that a full factorial computation is not feasible as the number 4096 of required simulation runs for a full factorial design exceeds all reasonable computation efforts.

However, it is neither necessary nor sensible to do a full factorial design, because, as the prob-

Implant			NMOS3V		PMOS3V	NMOS5V		PMOS5V
N-Well	P-Well	V_{TH}	I_{DS}	V_{TH}	I_{DS}	I_{DS}	V_{TH}	I_{DS}
1	1	1	0%	0%	0%	0%	0%	0%
1	1	0	0%	0%	-52%	0%	0%	-61%
1	1	2	0%	0%	66%	0%	0%	110%
1	0	1	6%	-9%	0%	12%	-20%	0%
1	2	1	-2%	9%	0%	-11%	24%	0%
0	1	1	0%	0%	15%	0%	0%	39%
0	2	1	-2%	9%	15%	-11%	24%	39%
2	1	1	0%	0%	-17%	0%	0%	-17%
2	0	1	6%	-9%	-17%	12%	-20%	-17%

Table 6.4: Calculated differences of the selected parameters to the nominal
implant set (1,1,1) in percent

ability for a process incident is rather small, the probability that multiple implant steps have been misprocessed is vanishing. However, it is not sufficient to calculate only the 32 situations for each single implant step being skipped or doubly processed, because scenarios where a wrong implant recipe is used lead to situations where one implant is missing and another one doubled. Hence, the possible combinations of missed and double implants have to be selected carefully. So, as can be seen from Table 1, out of the possible 27 combinations only 1 (process of record) + 3 times 2 (missing and double implant each) + 2 (swapped P- and N-Well implants) = 9 combinations remain that make sense.

Furthermore, one can distinguish between different implant "classes" which affect only certain electrical parameters. E.g., incidents related to the standard polysilicon resistor implant (poly 2 implant) can be easily detected by measurement of the polysilicon resistance. Hence, only three TCAD calculations need to be performed to cover possible incidents at this particular implant step.

This leads to the general requirement that some efforts are needed to identify an appropriate set of electrical parameters which will give an unambiguous indication of the "culprit" implant.

6.5.4 Simulation Results

The parameter values were extracted from combined process- and device simulations with the Synopsys software suite. Each step of the process flow relevant for the device structures was taken into account for the process simulation. After coming up with a device structure as shown in Figure 3.9 a device simulation was performed to obtain device characteristics like saturation current or NMOS threshold. The identical parameter extraction algorithms as in actual electrical tests were applied to enable a comparison to measured values for calibration of the simulation. Finally, the relative deviation of the nominal electrical parameters was calculated the results of which are shown in Table 6.4. These results show clearly the power of the proposed method for identifying root causes for wafer misprocessing. By choosing the driving capability and the threshold voltage of two different types of NMOS and the driving capability of two different types of PMOS transistors an unabigous set of electrical parameters was obtained. The percentage values in the table are relative changes of the parameters compared to the typical situation indicated in the first line of the table.

6.5.5 Conclusions

It has been shown that the proposed method has the power to identify root causes for wafer misprocessing quickly. Before the development of this method, it took valuable time of PCM data analysis by an experienced device engineer to find switched PLDD and NLDD reticles as the root cause for a misprocessed lot.

As these kinds of implant misprocessing incidents are rare, the system has to be understood as a preventive method to react to such problems as quickly as possible. It can save both, expensive engineering resources and additional measurements. Furthermore, the system can be used to rule out a number of speculations by simply trying them out with simulation and compare the "fingerprint" of their results.

'...Unter allen meinen Patienten jenseits der Lebensmitte, das heisst jenseits 35, ist nicht ein einziger, dessen endgültiges Problem nicht das der religiösen Einstellung wäre. [201]'
(I have treated many hundreds of patients. Among those in the second half of life - that is to say, over 35 - there has not been one whose problem in the last resort was not that of finding a religious outlook on life.)

Carl G. Jung

Chapter 7

Summary and Outlook

Implementing a framework for the integration of TCAD with the actual fabrication process results in multiple impacts on the strategic position of TCAD in a semiconductor fabrication environment. Historically TCAD was only applied on single device structures and only during process development to gain better insight into the physics behind devices [202]. Additionally, information on physical quantities which are difficult to obtain experimentally was gained. By automated integration of the TCAD framework over the whole work flow of semiconductor circuit fabrication many additional application fields can be addressed, as shown by this work. The setup of new processes (or the transfer of existing technologies) is speeded up dramatically. The human induced errors are consecutely reduced. The number of, at least passive, users of TCAD in a semiconductor company grows from a handful engineers to the entire engineering and production team. This results also in a much better utilization of the resources spent in TCAD (software license costs, work efficiency of TCAD engineers, computer hardware etc.). The gap in technical information between the top management and the "engineer in the production line" is made smaller. This aspect should not be underestimated in the field of semiconductor industry because due to the high complexity of integrated circuit fabrication, any closed documentation of the processes is of inevitable value.

However some open questions remain. The integration of etching and deposition recipies via automatic conversion is still on the level of transferring etch and deposition rates. The lack of generic equipment simulators for etching and deposition leads to additional effort in calibrating these steps in the TCAD simulation. Furthermore, there is still no fully automated approach to generate SPICE models from measurement or simulation data without user interaction. This leads to a significant amount of resource allocation at every additional model interaction. Finally, package related effects (thermal and electromagnetical) are not included on a routine basis yet. Since there exists a strong trend to convergence of different technologies (RF, MEMS, sensors, optical etc.), system on a chip (SOC) solutions will play a sigificant role in the future. Therefore not only the small silicon die, but the overall system consisting of die, bond wires, lead frame, and package body has to be taken into account as a whole.

SUMMARY AND OUTLOOK

Appendix A

Basis of the GAUSSIAN Normal Distribution Function

A.1 The GAUSSIAN Normal Distribution

A normal distribution in a variate[1] X with mean μ and variance σ^2 is a statistic distribution with the probability function

$$P(x) = \frac{1}{\sigma\sqrt{2\pi}} e^{-\frac{(x-\mu)^2}{2\sigma^2}} \tag{A.1}$$

on the domain $x \in (-\infty, \infty)$.

DEMOIVRE developed the normal distribution as an approximation to the binomial distribution, and it was subsequently used by LAPLACE in 1783 to study measurement errors and by GAUSS in 1809 in the analysis of astronomical data.

The so-called "standard normal distribution" is given by taking $\mu = 0$ and $\sigma^2 = 1$ in a general normal distribution. An arbitrary normal distribution can be converted to a standard normal distribution by changing variables to $Z \equiv \frac{X-\mu}{\sigma}$, so $dz = \frac{dx}{\sigma}$, yielding

$$P(x)dx = \frac{1}{\sqrt{2\pi}} e^{-\frac{z^2}{2}} dz \tag{A.2}$$

The normal distribution function $\phi(z)$ gives the probability that a standard normal variate assumes a value in the interval $[0, z]$,

$$\phi(z) \equiv \frac{1}{\sqrt{2\pi}} \int_0^z e^{-\frac{x^2}{2}} dx = \frac{1}{2} \operatorname{erf}\left(\frac{z}{\sqrt{2}}\right) \tag{A.3}$$

where $\operatorname{erf}(z)$ is the error function. Neither $\phi(z)$ nor $\operatorname{erf}(z)$ can be expressed in terms of finite additions, subtractions, multiplications, and root extractions, and so both must be either computed numerically or otherwise approximated.

---

[1]A variate is a generalization of the concept of a random variable that is defined without reference to a particular type of probabilistic experiment. It is defined as the set of all random variables that obey a given probabilistic law. It is common practice to denote a variate with a capital letter (most commonly X).

The normal distribution is the limiting case of a discrete binomial distribution $P_P(n|N)$ as the sample size N becomes large, in which case $P_P(n|N)$ is normal with mean and variance

$$\mu = Np \tag{A.4}$$
$$\sigma^2 = Npq \tag{A.5}$$

with $q \equiv 1 - p$.
The distribution $P(x)$ is properly normalized since

$$\int_{-\infty}^{\infty} P(x)dx = 1 \tag{A.6}$$

The cumulative distribution function which gives the probability that a variate will assume a value $\leq x$, is then the integral of the normal distribution

$$
\begin{aligned}
D(x) &\equiv \int_{-\infty}^{x} P(x')dx' \\
&= \frac{1}{\sigma\sqrt{2\pi}} \int_{-\infty}^{x} e^{-\frac{(x'-\mu)^2}{2\sigma^2}} dx' \\
&= \frac{1}{2} \left[1 + \operatorname{erf}\left(\frac{x-\mu}{\sigma\sqrt{2}}\right) \right]
\end{aligned}
\tag{A.7}
$$

The normal distribution function is obviously symmetric about $x = \mu$

$$P(x - \mu) = P(-x - \mu) \tag{A.8}$$

and its maximum value is situated at $x = \mu$

$$P_{max} = P(x = \mu) = \frac{1}{\sigma\sqrt{2\pi}} \tag{A.9}$$

Normal distributions have many convenient properties, so random variates with unknown distributions are often assumed to be normal, especially in physics and astronomy. Although this can be a dangerous assumption, it is often a good approximation due to a surprising result known as the central limit theorem (see next section).
Among the amazing properties of the normal distribution are that the normal sum distribution and normal difference distribution obtained by respectively adding and subtracting variates X and Y from two independent normal distributions with arbitrary means and variances are also normal. The normal ratio distribution obtained from $\frac{X}{Y}$ has a Cauchy distribution.
The unbiased estimator[2] for the variance of a normal distribution is given by

$$\sigma^2 = \frac{N}{N-1} s^2 \tag{A.10}$$

[2] An estimator $\hat{\theta}$ is an unbiased estimator of θ if $\langle \hat{\theta} \rangle = \theta$.

where

$$s^2 \equiv \frac{1}{N} \sum_{i=1}^{N} (x_i - \bar{X})^2. \tag{A.11}$$

The characteristic function for the normal distribution is

$$\phi(t) = e^{\imath m t - \sigma^2 \frac{t^2}{2}}, \tag{A.12}$$

and the moment-generating function is

$$
\begin{aligned}
M(t) &= \langle e^{tx} \rangle \\
&= \int_{-\infty}^{\infty} \frac{e^{tx}}{\sigma\sqrt{2\pi}} e^{-\frac{(x-\mu)^2}{2\sigma^2}} dx \\
&= e^{\mu t + \sigma^2 \frac{t^2}{2}}, \tag{A.13}
\end{aligned}
$$

so

$$
\begin{aligned}
M'(t) &= (\mu + \sigma^2 t) e^{\mu t + \sigma^2 \frac{t^2}{2}} \\
M''(t) &= \left[\sigma^2 + (\mu + t\sigma^2)^2\right] e^{\mu t + \sigma^2 \frac{t^2}{2}} \tag{A.14}
\end{aligned}
$$

A.2   The Central Limit Theorem

A.2.1   Theorem

Let $X_1, X_2, \ldots, X_N$ be a set of N independent random variates and X_i have an *arbitrary* probability distribution $P(x_1, \ldots, x_N)$ with mean μ_i and a finite variance σ_i^2.
Two variates A and B are statistically independent if the conditional probability $P(A|B) = \frac{P(A \cap B)}{P(B)}$ (probability of an event A assuming that B has occurred) of A given B satisfies

$$P(A|B) = P(A) \tag{A.15}$$

in which case the probability of A and B is just

$$P(AB) = P(A \cap B) = P(A)P(B) \tag{A.16}$$

Similarly, n events $A_1, A_2, \ldots, A_n$ are independent if

$$P\left(\bigcap_{i=1}^{n} A_i\right) = \prod_{i=1}^{n} P(A_i) \tag{A.17}$$

Then the normal form variate

$$X_{norm} = \frac{\sum_{i=1}^{N} x_i - \sum_{i=1}^{N} \mu_i}{\sqrt{\sum_{i=1}^{N} \sigma_i^2}} \tag{A.18}$$

has a limiting cumulative distribution function which approaches a normal distribution. Under additional conditions on the distribution of the variates, the probability density itself is also normal with mean $\mu = 0$ and variance $\sigma^2 = 1$. If conversion to normal form is not performed, then the variate

$$X \equiv \frac{1}{N} \sum_{i=1}^{N} x_i \tag{A.19}$$

is normally distributed with $\mu_X = \mu_x$ and $\sigma_X = \frac{\sigma_x}{\sqrt{N}}$.

A.2.2 Proof

Consider the inverse FOURIER transform of $P_X(f)$.

$$
\begin{aligned}
\mathcal{F}_f^{-1}[P_X(f)](x) &\equiv \int_{-\infty}^{\infty} e^{2\pi i f X} P(X) dX \\
&= \int_{-\infty}^{\infty} \sum_{n=0}^{\infty} \frac{(2\pi i f X)^n}{n!} P(X) dX \\
&= \sum_{x=0}^{\infty} \frac{(2\pi i f)^n}{n!} \int_{-\infty}^{\infty} X^n P(X) dX \\
&= \sum_{x=0}^{\infty} \frac{(2\pi i f)^n}{n!} \langle X^n \rangle. \tag{A.20}
\end{aligned}
$$

$$\tag{A.21}$$

Now write

$$\langle X^n \rangle = \langle N^{-n}(x_1 + x_2 + \ldots + x_N)^n \rangle = \tag{A.22}$$

$$\int_{-\infty}^{\infty} N^{-n}(x_1 + \ldots + x_N)^n P(x_1) \cdots P(x_N) dx_1 \cdots dx_N, \tag{A.23}$$

so we have

$$
\begin{aligned}
\mathcal{F}_f^{-1}[P_X(f)](x) &= \sum_{x=0}^{\infty} \frac{(2\pi i f)^2}{n!} \langle X^n \rangle \\
&= \sum_{x=0}^{\infty} \frac{(2\pi i f)^2}{n!} \int_{-\infty}^{\infty} N^{-n}(x_1 + \ldots + x_N)^n \times P(x_1) \cdots P(x_N) dx_1 \cdots dx_N \\
&= \int_{-\infty}^{\infty} \sum_{n=0}^{\infty} \left[\frac{2\pi i f(x_1 + \ldots + x_N)}{N} \right]^n \frac{1}{n!} P(x_1) \cdots P(x_N) dx_1 \cdots dx_N \\
&= \int_{-\infty}^{\infty} e^{\frac{2\pi i f(x_1 + \cdots + x_N)}{N}} P(x_1) \cdots P(x_N) dx_1 \cdots dx_N \\
&= \left[\int_{-\infty}^{\infty} e^{\frac{2\pi i f x_1}{N}} P(x_1) dx_1 \right] \times \cdots \times \left[\int_{-\infty}^{\infty} e^{\frac{2\pi i f x_N}{N}} P(x_N) dx_N \right] \\
&= \left[\int_{-\infty}^{\infty} e^{\frac{2\pi i f x}{N}} P(x) dx \right]^N \\
&= \left\{ \int_{-\infty}^{\infty} \left[1 + \left(\frac{2\pi i f}{N} \right) x + \frac{1}{2} \left(\frac{2\pi i f}{N} \right)^2 x^2 + \ldots \right] P(x) dx \right\}^N \\
&= \left[1 + \frac{2\pi i f}{N} \langle x \rangle - \frac{(2\pi f)^2}{2N^2} \langle x^2 \rangle + \mathcal{O}(N^{-3}) \right]^N \\
&= e^{N \ln \left[1 + \frac{2\pi i f}{N} \langle x \rangle - \frac{(2\pi f)^2}{2N^2} \langle x^2 \rangle + \mathcal{O}(N^{-3}) \right]}.
\end{aligned}
\tag{A.24}
$$

Now expand

$$
\ln(1 + x) = x - \frac{1}{2}x^2 + \frac{1}{3}x^3 + \ldots,
\tag{A.25}
$$

so

$$
\begin{aligned}
\mathcal{F}_f^{-1}[P_X(f)](x) &\approx e^{N \left[\frac{2\pi i f}{N} \langle x \rangle - \frac{(2\pi f)^2}{2N^2} \langle x^2 \rangle + \frac{1}{2} \frac{(2\pi i f)^2}{N^2} \langle x \rangle^2 + \mathcal{O}(N^{-3}) \right]} \\
&= e^{2\pi i f \langle x \rangle - \frac{(2\pi f)^2(\langle x^2 \rangle - \langle x \rangle^2)}{2N} + \mathcal{O}(N^{-2})} \\
&\approx e^{2\pi i f \mu_x - \frac{(2\pi f)^2 \sigma_x^2}{2N}},
\end{aligned}
\tag{A.26}
$$

since

$$
\mu_x \equiv \langle x \rangle
\tag{A.27}
$$

$$
\sigma_x^2 \equiv \langle x^2 \rangle - \langle x \rangle^2
\tag{A.28}
$$

Taking the FOURIER transform

$$P_X \equiv \int\limits_{-\infty}^{\infty} e^{-2\pi i f x} \mathcal{F}^{-1}\left[P_X(f)\right] df$$

$$= \int\limits_{-\infty}^{\infty} e^{2\pi i f(\mu_x - x) - (2\pi f)^2 \frac{\sigma_x^2}{2N}} df. \tag{A.29}$$

This is of the form

$$\int\limits_{-\infty}^{\infty} e^{iaf - bf^2} df, \tag{A.30}$$

where $a \equiv 2\pi(\mu_x - x)$ and $b \equiv \frac{(2\pi\sigma_x)^2}{2N}$. This integral yields

$$\int\limits_{-\infty}^{\infty} e^{iaf - bf^2} df = e^{-\frac{a^2}{4b}} \sqrt{\frac{\pi}{b}} \tag{A.31}$$

Therefore

$$P_X = \sqrt{\frac{\pi}{\frac{(2\pi\sigma_x)^2}{2N}}} e^{\frac{-[2\pi(\mu_x - x)]^2}{4\frac{(2\pi\sigma_x)^2}{2N}}}$$

$$= \sqrt{\frac{2\pi N}{4\pi^2 \sigma_x^2}} e^{-\frac{4\pi^2(\mu_x - x)^2 2N}{4\cdot 4\pi^2 \sigma_x^2}}$$

$$= \frac{\sqrt{N}}{\sigma_x \sqrt{2\pi}} e^{-\frac{(\mu_x - x)^2 N}{2\sigma_x^2}}. \tag{A.32}$$

But $\sigma_X = \frac{\sigma_x}{\sqrt{N}}$ and $\mu_X = \mu_x$, so

$$P_X = \frac{1}{\sigma_X \sqrt{2\pi}} e^{-\frac{(\mu_X - x)^2}{2\sigma_X^2}} \tag{A.33}$$

The "fuzzy" central limit theorem says that data which are influenced by many small and unrelated random effects are approximately normally distributed.

Appendix B

Layout Data Formats

B.1   Caltech Intermediate Format (CIF)

Caltech Intermediate Format (CIF) is a recent form for the description of integrated circuits. Created by the university community, CIF has provided a common database structure for the integration of many research tools. CIF provides a limited set of graphics primitives that are useful for describing the two-dimensional shapes on the different layers of a chip. The format allows hierarchical description which makes the representation concise. In addition, it is a terse but human-readable text format. CIF is therefore a concise and powerful descriptive form for VLSI geometry.

Each statement in CIF consists of a keyword or letter followed by parameters and terminated with a semicolon. Spaces must separate the parameters but there are no restrictions on the number of statements per line or of the particular columns of any field. Comments can be inserted anywhere by enclosing them in parenthesis.

There are only a few CIF statements and they fall into one of two categories: geometry or control. The geometry statements are: LAYER to switch mask layers, BOX to draw a rectangle, WIRE to draw a path, ROUNDFLASH to draw a circle, POLYGON to draw a figure, and CALL to invoke a subroutine of other geometry statements. The control statements are DS to start the definition of a subroutine, DF to finish the definition of a subroutine, DD to delete the definition of subroutines, 0 through 9 to include additional user-specified information, and END to terminate a CIF file. All of these keywords are usually abbreviated to one or two letters that are unique.

B.1.1   Geometry

The LAYER statement (or the letter L) sets the mask layer to be used for the subsequent geometry statements. Following the LAYER keyword comes a single layer-name parameter. For example, the command:

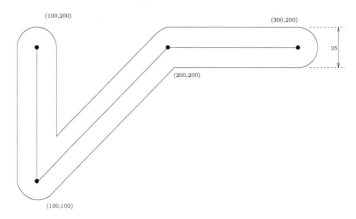

Figure B.1: A sample CIF "wire" statement. The statement is: `W25 100 200` `100 100 200 200 300 200;`

```
L NC;
```

sets the layer to NC, which often stands for the NMOS contact hole.

The BOX statement (or the letter B) is the most commonly used way of specifying geometry. It describes a rectangle by giving its length, width, center position, and an optional rotation. The format is as follows:

B *length width xpos ypos [rotation]* ;

Without the *rotation* field, the four numbers specify a box the center of which is at (*xpos, ypos*) and is *length* across in x and *width* tall in y. All numbers in CIF are integers that refer to centimicrons of distance, unless subroutine scaling is specified (described later). The optional *rotation* field contains two numbers that define a vector endpoint starting at the origin. The default value of this field is (1, 0), which is a right-pointing vector. Thus the *rotation* clause 10 5 defines a 30-degree counterclockwise rotation from the normal. Similarly, 10 -10 will rotate clockwise by 45 degrees. Note that the magnitude of this rotation vector has no meaning.

The WIRE statement (or the letter W) is used to construct a path that runs between a set of points. The path can have a nonzero width and has rounded corners. After the WIRE keyword comes the width value and then an arbitrary number of coordinate pairs that describe the endpoints. Figure B.1 shows a sample wire. Note that the endpoint and corner rounding are implicitly handled.

The ROUNDFLASH statement (or the letter R) draws a filled circle, given the diameter and the center coordinate. For example, the statement:

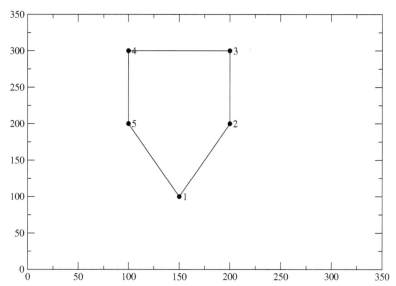

Figure B.2: A sample CIF "polygon" statement. The statement is: P 150 100 200 200 200 300 100 300 100 200;

```
R 20 30 40;
```

will draw a circle that has a radius of 10 (diameter of 20), centered at (30, 40). The POLYGON statement (or the letter P) takes a series of coordinate pairs and draws a filled polygon from them. Since filled polygons must be closed, the first and last coordinate points are implicitly connected and need not be the same. Polygons can be arbitrarily complex, including concavity and self-intersection. Figure B.2 illustrates a polygon statement.

B.1.2 Hierarchy

The CALL statement (or the letter C) invokes a collection of other statements that have been packaged with DS and DF. All subroutines are given numbers when they are defined and these numbers are used in the CALL to identify them. If, for example, a LAYER statement and a BOX statement are packaged into subroutine 4, then the statement:

```
C 4;
```

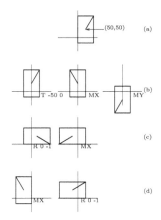

Figure B.3: The transformations of a CIF "call": (a) Subroutine 10: BOX 100 200 50 50; WIRE 10 50 50 100 150; (b) Invocation: C 10 T -50 0 MX MY; (c) Invocation: C 10 R 0 -1 MX; (d) Invocation: C 10 MX R 0 -1;

will cause the box to be drawn on that layer. In addition to simply invoking the subroutine, a CALL statement can include transformations to affect the geometry inside the subroutine. Three transformations can be applied to a subroutine in CIF: translation, rotation, and mirroring. Translation is specified as the letter T followed by an x, y offset. These offsets will be added to all coordinates in the subroutine, to translate its graphics across the mask. Rotation is specified as the letter R followed by an x, y vector endpoint that, much like the rotation clause in the BOX statement, defines a line to the origin. The unrotated line has the endpoint (1, 0), which points to the right. Mirroring is available in two forms: MX to mirror in x and MY to mirror in Y. Mirroring is a bit confusing, because MX causes a negation of the x coordinate, which effectively mirrors about the y axis! Any number of transformations can be applied to an object and their listed order is the sequence that will be used to apply them. Figure B.3 shows some examples, illustrating the importance of ordering the transformations (notice that Figure B.3(c) and Figure B.3(d) produce different results by rearranging the transformations).

Defining subroutines for use in a CALL statement is quite simple. The statements to be packaged are enclosed between DS (definition start) and DF (definition finish) statements. Arguments to the DS statement are the subroutine number and a subroutine scaling factor. There are no arguments to the DF statement. The scaling factor for a subroutine consists of a numerator followed by a denominator that will be applied to all values inside the subroutine. This scaling allows large numbers to be expressed with fewer digits and allows ease of rescaling a design. The scale factor cannot be changed for each invocation of the subroutine since it is applied to the definition. As an example, the subroutine of Figure B.3 can be described formally as follows:

```
DS 10 20 2;
```

```
      B10 20 5 5;
      W1 5 5 10 15;
DF;
```

Note that the scale factor is 20/2, which allows the trailing zero to be dropped from all values inside the subroutine. Arbitrary depth of hierarchy is allowed in CIF subroutines. Forward references are allowed provided that a subroutine is defined before it is used. Thus the sequence:

```
   DS 10;


      . . .
      C 11;
   DF;
   C 10;
```

is legal, but the sequence:

```
   C 11;
   DS 11;


      . . .


   DF;
```

is not. This is because the actual invocation of subroutine 11 does not occur until after its definition in the first example.

B.1.3   Control

CIF subroutines can be overwritten by deleting them and then redefining them. The DD statement (delete definition) takes a single parameter and deletes every subroutine that has a number greater than or equal to this value. The statement is useful when merging multiple CIF files because designs can be defined, invoked, and deleted without causing naming conflicts. However, it is not recommended for general use by CAD systems. Extensions to CIF can be done with the numeric statements 0 through 9. Although not officially part of CIF, certain conventions have evolved for the use of these extensions (see Table B.1). The final statement in a CIF file is the END statement (or the letter E). It takes no parameters and typically does not include a semicolon.

0 x y layer N name;	Set named node on specified layer and position
0V x1 y1 x2 y2 ... xn yn;	Draw vectors
2A "msg" *T x y;*	Place message above specified location
2B "msg" *T x y;*	Place message below specified location
2C "msg" *T x y;*	Place message centered at specified location
2L "msg" *T x y;*	Place message left of specified location
2R "msg" *T x y;*	Place message right of specified location
4A *lowx lowy highx highy;*	Declare cell boundary
4B *instancename;*	Attach instance name to cell
4N *signalname x y;*	Labels a signal at a location
9 *cellname;*	Declare cell name
91 *instancename;*	Attach instance name to cell
94 *label x y;*	Place label in specified location
95 *label length width x y;*	Place label in specified area

Table B.1: Typical user extensions to CIF

B.2   Calma GDS II stream format (GDSII)

In the design of integrated circuits, the most popular format for interchange is the Calma GDS II stream format (GDS II is a trademark of Calma Company, a wholly owned subsidiary of General Electric Company, U.S.A.). For many years, this format was the only one of its kind and many other vendors accepted it in their systems. Although Calma has updated the format as their CAD systems have developed, they have maintained backward compatibility so that no GDS II files become obsolete. This is important because GDS II is a binary format that makes assumptions about integer and floating-point representations. A GDS II circuit description is a collection of cells that may contain geometry or other cell references. These cells, called **structures** in GDS II parlance, have alphanumeric names up to 32 characters long. A library of these structures is contained in a file that consists of a library header, a sequence of structures, and a library tail. Each structure in the sequence consists of a structure header, a sequence of **elements**, and a structure tail. There are seven kinds of elements: **boundary** defines a filled polygon, **path** defines a wire, **structure reference** invokes a subcell, **array reference** invokes an array of subcells, **text** is for documentation, **node** defines an electrical path, and **box** places rectangular geometry.

B.2.1   Record Format

In order to understand the precise format of the above GDS II components, it is first necessary to describe the general record format. Each GDS II record has a 4-byte header that specifies the record size and function. The first 2 bytes form a 16-bit integer that contains the record length in bytes. This length includes the 4-byte header and must always be an even number. The end of a record can contain a single null byte if the record contents is an odd number of bytes long. The third byte of the header contains the type of the record and the fourth byte contains the type of the data. Since the data type is constant for each record type, this 2-byte field defines

File Header Records:	Bytes 3 and 4	Parameter Type
HEADER	0002	2-byte integer
BGNLIB	0102	12 2-byte integers
LIBNAME	0206	ASCII string
REFLIBS	1F06	2 45-character ASCII strings
FONTS	2006	4 44-character ASCII strings
ATTRTABLE	2306	44-character ASCII string
GENERATIONS	2202	2-byte integer
FORMAT	3602	2-byte integer
MASK	3706	ASCII string
ENDMASKS	3800	No data
UNITS	0305	2 8-byte floats
File Tail Records:	**Bytes 3 and 4**	**Parameter Type**
ENDLIB	0400	No data
Structure Header Records:	**Bytes 3 and 4**	**Parameter Type**
BGNSTR	0502	12 2-byte integers
STRNAME	0606	Up to 32-characters ASCII string
Structure Tail Records:	**Bytes 3 and 4**	**Parameter Type**
ENDSTR	0700	No data

Table B.2: GDS II header record types

the possible records as shown in Table B.2 and Table B.3.

B.2.2 Library Head and Tail

A GDS II file header always begins with a HEADER record the parameter of which contains the GDS II version number used to write the file. For example, the bytes 0, 6, 0, 2, 0, 1 at the start of the file constitute the header record for a version-1 file. Following the HEADER comes a BGNLIB record that contains the date of the last modification and the date of the last access to the file. Dates take six 2-byte integers to store the year, month, day, hour, minute, and second. The third record of a file is the LIBNAME, which identifies the name of this library file. For example, the bytes 0, 8, 2, 6, "C", "H", "I", "P" define a library named "CHIP." Following the LIBNAME record there may be any of the optional header records: REFLIBS to name up to two reference libraries, FONTS to name up to four character fonts, ATTRTABLE to name an attribute file, GENERATIONS to indicate the number of old file copies to keep, and FORMAT to indicate the nature of this file. The strings in the REFLIBS, FONTS, and ATTRTABLE records must be the specified length, padded with zero bytes. The parameter to FORMAT has the value 0 for an archived file and the value 1 for a filtered file. Filtered files contain only a subset of the mask layers and that subset is described with one or more MASK records followed by an ENDMASK record. The string parameter in a MASK record names layers and sequences of layers; for example, "1 3 5-7." The final record of a file header must be the UNITS record. The parameters to this record contain the number of user units per database unit (typically less than 1 to allow granularity of user specification) and the number of meters per database unit

Element Header Records:	Bytes 3 and 4	Parameter Type
BOUNDARY	0800	No data
PATH	0900	No data
SREF	0A00	No data
AREF	0B00	No data
TEXT	0C00	No data
NODE	1500	No data
BOX	2D00	No data
Element Contents Records:	**Bytes 3 and 4**	**Parameter Type**
ELFLAGS	2601	2-byte integer
PLEX	2F03	4-byte integer
LAYER	0D02	2-byte integers
DATATYPE	0E02	2-byte integer
XY	1003	Up to 200 4-byte integer pairs
PATHTYPE	2102	2-byte integer
WIDTH	0F03	4-byte integer
SNAME	1206	Up to 32-character ASCII string
STRANS	1A01	2-byte integer
MAG	1B05	8-byte float
ANGLE	1C05	8-byte float
COLROW	1302	2 2-byte integers
TEXTTYPE	1602	2-byte integer
PRESENTATION	1701	2-byte integer
ASCII STRING	1906	Up to 512-character string
NODETYPE	2A02	2-byte integer
BOXTYPE	2E02	2-byte integer

Table B.3: GDS II element record types

(typically much less than 1 for IC specifications). Eight-byte floating-point numbers have a sign at the top of the first byte, a 7-bit exponent in the rest of that byte, and 7 more bytes that compose a mantissa (all to the right of an implied decimal point). The exponent is a factor of 16 in excess-64 notation (that is, the mantissa is multiplied by 16 raised to the true value of the exponent, where the true value is its integer representation minus 64). Following the file header records come the structure records. After the last structure has been defined, the file terminates with a simple ENDLIB record. Note that there is no provision for the specification of a root structure to define a circuit; this must be tracked by the designer.

B.2.3 Structure Head and Tail

Each structure has two header records and one tail record that sandwich an arbitrary list of elements. The first structure header is the BGNSTR record, which contains the creation date and the last modification date. Following that is the STRNAME record, which names the structure using any alphabetic or numeric characters, the dollar sign, or the underscore. The structure is then open and any of the seven elements can be listed. The last record of a structure is the ENDSTR. Following it must be another BGNSTR or the end of the library, ENDLIB.

B.2.4 Boundary Element

The boundary element defines a filled polygon. It begins with a BOUNDARY record, has an optional ELFLAGS and PLEX record, and then has required LAYER, DATATYPE, and XY records. The ELFLAGS record which appears optionally in every element, has two flags in its parameter to indicate template data (if bit 16 is set) or external data (if bit 15 is set). This record should be ignored on input and excluded from output. Note that the GDS II integer has bit 1 in the leftmost or most significant position so these two flags are in the least significant bits. The PLEX record is also optional to every element and defines element structuring by aggregating those that have common plex numbers. Although a 4-byte integer is available for plex numbering, the high byte (first byte) is a flag that indicates the head of the plex if its least significant bit (bit 8) is set. The LAYER record is required to define which layer (numbered 0 to 63) is to be used for this boundary. The meaning of these layers is not defined rigorously and must be determined for each design environment and library. The DATATYPE record contains unimportant information and its argument should be zero. The XY record contains anywhere from four to 200 coordinate pairs that define the outline of the polygon. The number of points in this record is defined by the record length. Note that boundaries must be closed explicitly, so the first and last coordinate values must be the same.

B.2.5 Path Element

A path is an open figure with a nonzero width that is typically used to place wires. This element is initiated with a PATH record followed by the optional ELFLAGS and PLEX records. The LAYER record must follow to identify the desired path material. Also, a DATATYPE record must appear and an XY record to define the coordinates of the path. From two to 200 points may be given in a path. Prior to the XY record of a path specification there may be two optional records called PATHTYPE and WIDTH. The PATHTYPE record describes the nature of the path segment ends, according to its parameter value. If the value is 0, the segments will have square ends that terminate at the path vertices. The value 1 indicates rounded ends and the value 2 indicates square ends that overlap their vertices by one-half of their width. The width of the path is defined by the optional WIDTH record. If the width value is negative, then it will be independent of any structure scaling (from MAG records, see next section).

B.2.6 Structure Reference Element

Hierarchy is achieved by allowing structure references (instances) to appear in other structures. The SREF record indicates a structure reference and is followed by the optional ELFLAGS and PLEX records. The SNAME record then names the desired structure and an XY record contains a single coordinate to place this instance. It is allowed to reference structures that have not yet been defined with STRNAME. Prior to the XY record there may be optional transformation records. The STRANS record must appear first if structure transformations are desired. Its parameter has bit flags that indicate mirroring in x before rotation (if bit 1 is set), the use of absolute magnification (if bit 14 is set), and the use of absolute rotation (if bit 15 is set). The magnification and rotation amounts may then be specified in the optional MAG and ANGLE records. The rotation angle is in counterclockwise degrees.

B.2.7 Array of Structures Element

For convenience, an array of structure instances can be specified with the AREF record. Following the optional ELFLAGS and PLEX records comes the SNAME to identify the structure being arrayed. Next, the optional transformation records STRANS, MAG, and ANGLE give the orientation of the instances. A COLROW record must follow to specify the number of columns and the number of rows in the array. The final record is an XY with three points: the coordinate of the corner instance, the coordinate of the last instance in the columnar direction, and the coordinate of the last instance in the row direction. From this information, the amount of instance overlap or separation can be determined. Note that flipping arrays (in which alternating rows or columns are mirrored to abut along the same side) can be implemented with multiple arrays that are interlaced and spaced apart to describe alternating rows or columns.

B.2.8 Text Element

Messages can be included in a circuit with the TEXT record. The optional ELFLAGS and PLEX follow with the mandatory LAYER record after that. A TEXTTYPE record with a zero argument must then appear. An optional PRESENTATION record specifies the font in bits 11 and 12, the vertical presentation in bits 13 and 14 (0 for top, 1 for middle, 2 for bottom), and the horizontal presentation in bits 15 and 16 (0 for left, 1 for center, 2 for right). Optional PATHTYPE, WIDTH, STRANS, MAG, and ANGLE records may appear to affect the text. The last two records are required: an XY with a single coordinate to locate the text and a STRING record to specify the actual text.

B.2.9 Node Element

Electrical nets may be specified with the NODE record. The optional ELFLAGS and PLEX records follow and the required LAYER record is next. A NODETYPE record must appear with a zero argument, followed by an XY record with one to 50 points that identify coordinates on the electrical net. The information in this element is not graphical and does not affect the manufactured circuit. Rather, it is for other CAD systems that use topological information.

B.2.10 Box Element

The last element of a GDS II file is the box. Following the BOX record are the optional ELFLAGS and PLEX records, a mandatory LAYER record, a BOXTYPE record with a zero argument, and an XY record. The XY must contain five points that describe a closed, four-sided box. Unlike the boundary, this is not a filled figure. Therefore it cannot be used for IC geometry.

B.3   Electronic Design Interchange Format (EDIF)

The Electronic Design Interchange Format (EDIF) is a recent effort at capturing all aspects of VLSI design in a single representation. This is useful not only as a communications medium to manufacturing equipment, but also as an interchange format between CAD systems, since none of the high-level information is lost. EDIF is designed to be both easy to read by humans and easy to parse by machines. EDIF files resemble the LISP programming language because of the use of prefix notation enclosed in parentheses. For example, the CIF polygon:

```
P 100 100 150 200 200 100;
```

look like this in EDIF:

```
(polygon (point 100 100) (point 150 200) (point 200 100))
```

All EDIF statements consist of an open parenthesis, a keyword, some parameters, and a close parenthesis. The parameters can be other statements, which is what gives EDIF structure. Actually, an EDIF file contains only one statement:

```
(edif parameters)
```

where parameters are the described circuit. Not only does EDIF resemble LISP, but in its highest level it contains all of LISP and is an extension of this highly expressive language. However, in the interest of making parsing simple, there are three levels of EDIF, and lower levels are less powerful. Level 1, the intermediate level, allows variables to be used and cell definitions to be parameterized. EDIF level 0 has no programmability and requires constants in all statements. A LISP preprocessor can translate from EDIF levels 1 or 2 down to level 0, and any given level of EDIF can be read by a parser of a higher level. Since level 0 is all that is necessary for most interchange and all manufacturing specification, only that level will be discussed here. Also, some of the EDIF constructs that deal with simulation, routing, behavior, and other unusual specifications will not be covered in detail.

B.3.1 EDIF Structure

An EDIF file contains a set of **libraries**, each containing a set of **cells**. Each cell can be described with one or more **views** that show the cell in the form of a schematic, layout, behavioral specification, document, and more. Each view has both an **interface** and **contents** so that it is cleanly defined and can be linked to other views with a **view map**. Libraries may also contain **technology** information so that defaults can be given for behavior, graphics, and other attributes. The overall structure of an EDIF file looks like this:

```
(edif name


    (status information)
    (design where-to-find-them)
    (external reference-libraries)
    (library name


        (technology defaults)
        (cell name


            (viewmap map)
            (view type name


                (interface external)
                (contents internal)


            )


        )


    )


)
```

The **status** statement is used to track design progress and contains author names, modification dates, and program versions. Additional status statements may appear in each library, cell, and view. The **design** statement indicates where a completed design may be found by pointing to the top cell of a hierarchical description. The **external** statement names libraries that will be used but are not listed in this EDIF file. The **library, cell,** and **view** blocks can be repeated as necessary. There is also a comment statement that can be placed at the end of most blocks to add readability to the file. Here is an example of an EDIF file that further illustrates the outer level:

```
(edif my-design

    (status

        (edifversion 1 0 0)
        (ediflevel 0)
        (written

            (timestamp 2004 4 1 11 16 6)
            (accounting author "Rainer Minixhofer")
            (accounting location "Vienna")
            (accounting program "Virtuoso")
            (comment "timestamp contains year, month, day, hour,")
            (comment " minute, and second")

        )

    )
    (design test-chip

        (qualify test-library top-cell)
        (comment "look for top-cell in test-library")

    )
    (external pad-library pla-library)
    (library test-library

        (technology 0.35-micron-CMOS

            ...

        )
        (cell top-cell

            (viewmap ... )
            (view masklayout real-geometry
```

```
            (interface ... )
            (contents ... )

        )
        (view schematic more-abstract

            (interface ... )
            (contents ... )

        )

    )

  )

)
```

The `written` part of a `status` block may be repeated to show all authors and update events. Also note that the `qualify` statement which names a cell in a particular library, is generally useful and can appear anywhere that an isolated name may be ambiguous or undefined.

In the following sections, more information is given to describe the `contents`, `interface`, `viewmap`, and `technology` blocks.

B.3.2 Contents

The contents of a cell may be represented in a number of different ways depending on the type of data. Each representation is a different view, and multiple views can be used to define a circuit fully. EDIF accepts seven different view types: **netlist** for pure topology as is required by simulators, **schematic** for connected logic symbols, **symbolic** for more abstract connection designs, **mask layout** for the geometry of chip and board fabrication, **behavior** for functional description, **document** for general textual description, and **stranger** for any information that cannot fit into the other six view types. The statements allowed in the contents section vary with the view type (see Fig. D.1). The netlist, schematic, and symbolic views are essentially the same, because they describe circuit topology. The allowable statements in these views are the declarations *define, unused, global, rename,* and *instance*; the routing specifications *joined, mustjoin,* and *criticalsignal*; and the timing specifications *required* and *measured*. Schematic and symbolic views also allow the *annotate* and *wire* statements. The mask-layout view allows all of the declarations, some of the routing constructs, and the *figuregroup* statement for actual graphics. The behavior view allows only a few declarations and the *logicmodel* statement. The document view allows only the *instance* and *section* constructs. Finally, the stranger view allows everything but supports nothing. It should be avoided whenever possible.

	Netlist	Schematic	Symbolic	Mask Layout	Behavior	Document	Stranger
define	X	X	X	X	X		X
unused	X	X	X	X			X
global	X	X	X	X			X
rename	X	X	X	X	X		X
instance	X	X	X	X		X	X
joined	X	X	X	X			X
mustjoin	X	X	X	X			X
criticalsignal	X	X	X				X
required	X	X	X				X
measured	X	X	X				X
logicmodel					X		X
figuregroup				X			X
annotate		X	X				X
wire		X	X				X
section						X	X

Table B.4: Contents statements allowed in EDIF.

B.3.2.1 Declarations

Declarations establish the objects in a cell including signals, parts, and names. Internal signals are defined with the statement:

```
(define direction type names)
```

where *direction* is one of input, output, inout, local, or unspecified. Only local and unspecified signals have meaning in the contents section of a view; the others are used when this statement appears in an interface section. The *type* of the declaration can be port, signal, or figuregroup, where port is for the interface section, signal is for the contents section, and figuregroup is for the technology section of a library. The *names* being declared can be given as a single name or as a list of names aggregated with the multiple clause. In addition to all these declaration options, it is possible to define arrays by having any name be the construct:

```
(arraydefinition name dimensions)
```

These arrays can be indexed by using the construct:

```
(member name indexes)
```

Here are some examples of the **define** statement:

```
(define input port Clk)
(define unspecified signal (multiple a b c))
(define local signal (arraydefinition i-memory 10 32))
```

After the declaration of signals, a number of other declarations can be made. The **unused** statement has the form:

```
(unused name)
```

and indicates that the defined name is not used in this cell view and should be allowed in order to pass any analysis that might find it to be an error. The **global** declaration has the form:

```
(global names)
```

and defines signals to be used inside the cell view and one level lower, in subcomponents that are placed inside the view. Where these subcomponents have ports that match globally declared names, they will be implicitly equated. Yet another declaration is **rename**, which can associate any EDIF name with an arbitrary string. This allows illegal EDIF naming conventions to be used, such as in this example:

```
(rename bwest ''B-west{ss}'')
```

B.3.2.2   Instances

Hierarchy is achieved by declaring an instance of a cell in the contents of another cell. The format of the instance statement is as follows:

```
(instance cell view name transform)
```

The name of the other cell is in cell and the particular view to use is in view. This allows the hierarchy to mix view types where it makes sense. A local name is given to this instance in name and an optional transformation is specified in transform, which looks like this:

```
(transform scale rotate translate)
```

where scale can be:

    (scalex *numerator denominator*)

or:

    (scaley *numerator denominator*)

translate has the form:

    (translate *x y*)

and rotate can be either:

    (rotation *counterclockwise-angle*)

or:

    (orientation *manhattan-orientation*)

where manhattan-orientation is one of R0, R90, R180, R270, MX, MY, MYR90, or MXR90. So, for example, the expression:

    (transform (scalex 3 10) MX (translate 5 15))

will scale the instance to three-tenths of its size, mirror it about the x axis (negate the y coordinate) and then translate it by 5 in x and 15 in y. Although any of the three transformation elements can be omitted, when present they must be given in the order shown, and are applied in that sequence. Unfortunately, versions of EDIF before 3.0 have no provision for non-Manhattan orientation because the **rotation** clause did not exist before then.

Arrays of instances can be described by including a step function in the translate part of the transform clause. This will indicate a series of translated locations for the instances. The format of this iteration is:

    (step *origin count increment*)

So, for example, the clause:

```
(step 7 3 5)
```

will place three instances translated by 7, 12, and 17 in whichever coordinate this appears. The rotation and scale factors will apply to every array element. Also, it is possible to use different instances in each array location by mentioning multiple cell names in the *instance* clause. For example,

```
(instance (multiple carry add)

   more-abstract add-chain

      (transform (translate 0 (step 0 16 10))))

)
```

will create an array of 16 instances stacked vertically that alternate between the "more-abstract" view of the "carry" cell and the "more-abstract" view of the "add" cell. This entire instance will be called "add-chain."

B.3.2.3 Routing and Simulation

To indicate connectivity, the `joined` construct identifies signals or ports that are connected. The `mustjoin` construct indicates that signals do not yet connect but should when routing takes place, and the **criticalsignal** construct establishes priorities for the routing. To illustrate further EDIF's expressive power in routing specification, there is a **weakjoined** construct that defines a set of joins, only one of which must be connected, and a **permutable** statement that declares sets of connection points to be interchangeable. These last two statements are found only in the interface section; however, none of the routing constructs will be described in detail here.

The final set of constructs in the netlist, schematic, and symbolic views are those concerned with timing. As an example of the level of specification available, the statement:

```
(required (delay (transition H L (minomax 10 20 30)) here there))
```

specifies that the required delay of a high-to-low transition between points "here" and "there" is between 10 and 30, with a nominal value of 20. The **measured** statement can be used in the same way to document actual timing. The `logicmodel` statement, found only in the behavior view, allows a detailed set of logic states and conditions that can control simulation and verification. The EDIF specification should be consulted for full details of these timing, behavior, and routing constructs citeGEIAWWW.

B.3.2.4   Geometry

In the mask-layout view, geometry can be specified with the figuregroup construct which looks like this:

```
(figuregroup groupname


    pathtype width color fillpattern borderpattern
    signals figures)
```

where the *groupname* refers to a **figuregroupdefault** clause in the technology section of this library (described later). This set of defaults is available so that the graphic characteristics *pathtype, width, color, fillpattern,* and *borderpattern* need not be explicitly mentioned in each **figuregroup** statement. These five graphic characteristics are therefore optional and have the following format:

```
(pathtype endtype cornertype)
(width distance)
(color red green blue)
(fillpattern width height pattern)
(borderpattern length pattern)
```

The **pathtype** describes how wire ends and corners will be drawn (either **extend,truncate**, or round). The **width** clause takes a single integer to be used as the width of the wire. The **color** clause takes three integers in the range of 0 to 100, which give intensity of red, green, and blue. The **fillpattern** clause gives a raster pattern that will be tessellated inside of the figure. Two integers specify the size of the pattern and a string of zeros and ones define the pattern. Finally, the **borderpattern** describes an edge texture by specifying a single integer for a pattern length followed by a pattern string that is repeated around the border of the figure. Here are examples of these **figuregroup** attributes:

```
(pathtype round round)
(width 200)
(color 0 0 100)
(fillpattern 4 4 "1010010110100101")
(borderpattern 2 "10")
```

Inside of a **figuregroup** statement, the actual geometry can be specified directly with the *figures* constructs or can be aggregated by signal with the *signals* construct, which has the form:

```
(signalgroup name figures)
```

The figures construct in a figuregroup can be either `polygon`, `shape`, `arc`, `rectangle`, `circle`, `path`, `dot`, or `annotate`. The polygon is of the form:

> (polygon *points*)

where each point has the form:

> (point *x* *y*)

A `shape` is the same as a `polygon` except that it can contain `point` or `arc` information, freely mixed. The `arc` has the form:

> (arc *start middle end*)

where these three points are the start point, any point along the arc, and the endpoint. The `rectangle` takes two points that are diagonally opposite each other and define a rectangle. A `circle` takes two points that are on opposite sides and therefore separated by the diameter. The `path` takes a set of points and uses the `width` and `pathtype` information to describe the geometry further. The dot construct takes a single point and draws it in a dimensionless way (it should not be used in actual fabrication specifications, but can be used for internal documentation of special positions).

B.3.2.5 Miscellaneous Statements

In schematic and symbolic views, the annotate clause may be used to add text to a drawing. The form:

> (annotate *text corner1 corner2 justify*)

will place text in the box defined by the two diagonally opposite corners, and justify the text according to one of nine options: `upperleft`, `uppercenter`, `upperright`, `centerleft`, `centercenter`, `centerright`, `lowerleft`, `lowercenter`, or `lowerright`. For example:

> (annotate "probe here" (point 50 50) (point 200 100) uppercenter)

will place the string "probe here" in the upper part of the box $50 \leq x \leq 200$ and $50 \leq y \leq 100$, centered.

Another construct allowed only in schematic and symbolic views is `wire`. This connects two ports with a wire that can be described graphically. For example:

```
(wire clock.in gated.timer

    (figuregroup metal

        (path (point 10 15) (point 20 15) (point 20 25))

    )

)
```

connects the two points on the metal layer.

The last contents statement to be mentioned is the section construct, which is found only in document views and can hierarchically describe chapters, sections, subsections, and so on. For example:

```
(section "Chapter 1"

    (section "Introduction"

        "This is a doctoral thesis"
        "To receive the doctoral degree."

    )

)
```

B.3.3   Interface

In addition to there being seven different ways of specifying the contents of a cell, there are the same seven views that apply to the interface of a cell. The interface section is the specification of how a cell interacts with its environment when used in a supercell. Unlike the contents views, the seven interface views are all essentially the same (see Table B.5). The netlist, schematic, symbolic, mask layout, behavior, and stranger views can all contain the same declarations: `define`, `rename`, `unused`, `portimplementation`, and `body`. They also allow the routing statements `joined`, `mustjoin`, `weakjoined`, and `permutable`, in addition to the simulation statements `timing` and `simulate`. The symbolic, mask-layout, and stranger views add the `arrayrelatedinfo` construct, which enables gate-array specification to be handled. The document view offers no constructs as this text rightly belongs in the contents section.

	Netlist	Schematic	Symbolic	Mask Layout	Behavior	Stranger
define	X	X	X	X	X	X
rename	X	X	X	X	X	X
unused	X	X	X	X	X	X
portimplementation	X	X	X	X	X	X
body	X	X	X	X	X	X
joined	X	X	X	X	X	X
mustjoin	X	X	X	X	X	X
weakjoined	X	X	X	X	X	X
permutable	X	X	X	X	X	X
timing	X	X	X	X	X	X
simulate	X	X	X	X	X	X
arrayrelatedinfo			X	X		X

Table B.5: Interface statements allowed in EDIF.

B.3.3.1 Ports and Bodies

The first interface statement to be discussed is portimplementation, which describes the ports and their associated components, graphics, timing, and other properties. Although ports can be declared with the define statement, portimplementation allows more information to be included in the declaration. The format is:

    (portimplementation portname figuregroups instances properties)

where the portname is the name of the port as it will be used in supercells. The figuregroups describe any graphics attached to the port, the instances specify any subcells that describe the port, and the properties may indicate power-consumption ratings. Ports that are further described by instances of other cells do not need figuregroups to define them, so much of the portimplementation statement is optional.

The body statement is used to describe the external or interfaced aspect of a cell. In mask-layout views, this can describe a protection frame for design-rule checking and compaction. In other views it is simply used to give an external appearance to instances of the cell. The format is:

    (body figuregroups instances)

where instances are subcells that can be used to describe the body.

B.3.3.2   Gate-Array and Behavioral Interface

The `arrayrelatedinfo` statement which is used in gate-array specification, is allowed only in symbolic, mask-layout, and stranger views. This can be used to declare the background array:

```
(arrayrelatedinfo basearray (socket info))
```

or the individual cells:

```
(arrayrelatedinfo arraysite (plug info))
```

or macros of cells:

```
(arrayrelatedinfo arraymacro (plug info))
```

These statements define a grid that can be connected in a rigid manner, specified by the plugs and sockets. Sockets define permissible connection options and plugs make these connections to give precise gate-array interface [203].

The final interface section constructs, which will not be described in detail, are `timing` and `simulate`. The `timing` statement gives port delays for various transitions, and gives stability requirements for the signal values. The `simulate` statement lists test data and expected results.

B.3.4   View Maps

To relate different views, a `viewmap` section can exist in each cell, which associates ports from different interface sections or instances from different contents sections. Port mapping is done with:

```
(portmap ports)
```

where the list of ports is of the form:

```
(qualify viewname portname)
```

Thus to equate port C of the mask-layout view with port D of the schematic view, the map would look like this:

```
(viewmap
```

```
(portmap

    (qualify real-geometry C)
    (qualify more-abstract D)

  )


)
```

Note that the *viewname* is the declared name given to the view.

To relate instances of a cell in different views, the same format applies except that a many-to-one mapping is allowed. For example,

```
(instancemap

    (qualify real-geometry pullup pulldown)
    (qualify more-abstract inverter)

  )
```

will map both the pullup and the pulldown in the mask-layout view to the inverter in the schematic view.

B.3.5 Technology

The technology section provides a background of information for the description of a library. Defaults can be set for other statements in the library, such as the **figuregroup**. Also, the real units of distance, time, power, and so on can be established. The technology section has the following format:

```
(technology name

    defines renames
    figuregroupdefaults
    numberdefinitions gridmaps
    simulation

  )
```

where *name* is an identifier for this technology. A set of **define** statements can be used to declare default figuregroups for various signal types and rename statements can be used to establish name bindings in the library. The **figuregroupdefault** statement takes a name and a list of **pathtype, width, color, fillpattern,** and **borderpattern** constructs to establish the defaults for subsequent **figuregroup** statements in the library. The **numberdefinition** statement is important because it sets the scale of all EDIF units as follows:

```
(numberdefinition SI

    (scale distance edif real)
    (scale time edif real)
    (scale capacitance edif real)
    (scale current edif real)
    (scale resistance edif real)
    (scale voltage edif real)
    (scale temperature edif real)


)
```

The name SI is a standard that should always appear unless an alternate set of unit values is being declared. Any of the **scale** clauses may be used to declare the number of units in the EDIF file that correspond with real units. Real units for distance are in meters, which means that the clause:

```
(scale distance 1000000 1)
```

causes one million EDIF units to be a meter (or one EDIF unit to be a micron). The real-time unit is the second, capacitance is in farads, current is in amperes, resistance is in ohms, voltage is in volts, and temperature is in degrees celsius.

The **gridmap** clause of the technology section can be used to declare nonuniform scaling in the x and y axes. For example,

```
(gridmap 3 4)
```

will set the x units to be three times the **numberdefinition** distance and the y coordinates to be four times that amount. This nonuniform scaling of all coordinates has limited application.

A final use of the technology section is for simulation defaults. As with all other simulation constructs, these will not be discussed here.

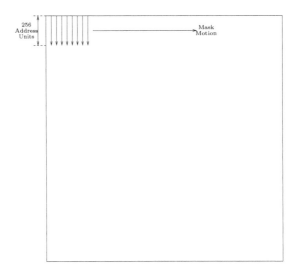

Figure B.4: EBES raster motion, actual mask making sweeps out 256 rows as it advances horizontally

B.4 Electron Beam Exposure System (EBES) Format

One of the most popular formats for producing integrated-circuit masks is the Electron Beam Exposure System (EBES), which can specify the very small features needed in high-density chips. Although the actual format varies with each different company's mask-making machine, an original standard, designed at Bell Laboratories, forms the basis on which extensions are made. This basic format is described here. The electron beam is controlled in a digital raster fashion, such that a line of 256 points (called **address units**) can be written at a time. By convention, this sweep is run vertically and the mask is moved horizontally to make the sweep cover the top 256 rows of the chip (see Figure B.4). It thus takes multiple passes across the width of the chip to write a complete pattern. When multiple copies of an IC die are being produced on a wafer, a single sweep of 256 rows is made on every die position before advancing to the next 256 rows. This means that the pattern file which is organized in 256 row **stripes**, needs to be read only once.

B.4.1 File Structure

The EBES file is organized about these 256 row stripes. The beginning of the file contains a START DRAWING command followed by the stripes and an END DRAWING command. Each stripe consists of a START STRIPE command, a series of **figure** commands, and an END STRIPE command. The figures can be either RECTANGLEs, PARALLELOGRAMs, or

TRAPEZOIDs.

EBES files are binary, with 16 bits per word. Commands can take any number of words but must be aggregated into 1024-word blocks. If a command is near the end of a block and would span into the next block, then an END OF BLOCK command must appear followed by pad data to the block end. All blocks must end with the END OF BLOCK command except for the last block, which ends with the END DRAWING command.

B.4.2 Control Commands

The START DRAWING command is the first in an EBES file and contains 16 words. The first word has a 2 in the high byte and a code in the low byte that gives the size of an address unit. The address-unit size will be 1 micron if the code is 0, one-half micron if the code is 3, and one-quarter micron if the code is 6. The second and third words of the START DRAWING command are the x and y size of the entire pattern. Words 4, 5, and 6 contain the EBES file creation date: Word 4 is the month, word 5 is the day, and word 6 is the year. For these fields, the high byte is the first digit and the low byte is the second digit in ASCII. Words 7 through 13 contain the name of this file by having a name length in word 7 (always the value 6) and by having the next six words (8 to 13) contain an ASCII string with the format "XXXXXXXXX.XX". Finally, words 14 through 16 describe the current mask layer by having word 14 contain the value 2 and words 15 and 16 contain four ASCII characters. After the START DRAWING command comes the first START STRIPE command. This command is one word with the value 7 in the low byte and the stripe number (from 1 to 255) in the high byte. Note that the word and byte sizes form restrictive limits on the overall pattern size. For larger dies to be made, multiple pattern files must be abutted. After the START STRIPE come the figures. When all figures in a stripe have been listed, an END STRIPE command appears. The END STRIPE is a single word with the value 8. At the end of a block is an END OF BLOCK command which is a single word with the value 9. At the end of the EBES file is an END DRAWING command which is a single word with the value 4.

B.4.3 Rectangles

The RECTANGLE is one of the figures allowed in a stripe. It contains four words that describe the width, height, and corner position within the stripe. The first word contains the value 16 in the low 6 bits and contains the rectangle height (minus one) in the high 10 bits. Since the rectangle must fit in a 256-tall stripe, this height can have only the values 0 to 255. The second word of the rectangle command is the width, the third word is the starting x position, and the fourth word is the starting y position. Again, the y position can range only from 0 to 255 since it must be within the stripe.

B.4.4 Parallelograms

The PARALLELOGRAM command is similar to the RECTANGLE command except that it has two more words to give a skew distance and extra precision bits (see Figure B.5). Word 1 has the value 17 in the low 6 bits and a value that is one less than the height in the top 10 bits.

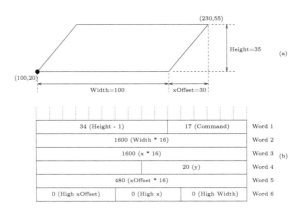

Figure B.5: EBES parallelogram example. (a) Figure (b) Record.

Words 2, 3 and 4 define the width, x position, and y position as in the RECTANGLE command, except words 2 and 3 are multiplied by 16. Word 5 contains the x offset between the bottom and the top (also multiplied by 16). The sixth word contains extra precision bits for the width (the low 5 bits), the x location (the next 5 bits), and the x offset (the high 6 bits). These extra precision bits are high bits for the three horizontal position values, and define 21- or 22-bit fields for each. The reason for this extra precision is not to extend the addressing range but to allow fractional coordinates: These three values are in 1/16 address units. Thus there is an implied decimal point 4 bits from the right on all these numbers.

B.4.5 Trapezoids

There are three types of TRAPEZOIDs, as illustrated by Figure B.6. Type 1 has a vertical right edge, Type 2 has a vertical left edge, and Type 3 has no vertical edges. The Type 1 and Type 2 TRAPEZOIDs can be described in exactly the same format as is used for the PARALLELOGRAM because they have the same fields: x, y, width, height, and x offset. Type 1 TRAPEZOIDs have the value 18 in the low 6 bits of the first word, and Type 2 TRAPEZOIDs have the value 19 in this field.

Type 3 TRAPEZOIDs require a seven-word description because of the extra x offset field. The first Word has a 20 in the low 6 bits and the height minus one in the top 10 bits. The second Word is the width, the third word is the x location, the fourth word is the y location, the fifth word is the first x offset, the sixth word is the second x offset, and the last word contains additional precision bits for the width (low 5 bits), the x location (next 5 bits), and the first x offset (top 6 bits). The extra precision bits for the second x offset are in the fourth word (the top 6 bits), which is otherwise used only for its low 8 bits to contain the y location.

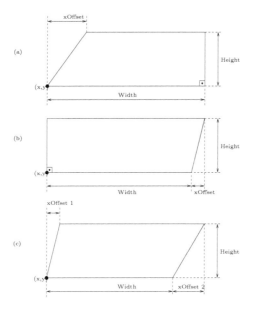

Figure B.6: **EBES trapezoid types:** (a) Type 1 has vertical edge on right (b) Type 2 has vertical edge on left (c) Type 3 has no vertical edges.

Appendix C

General Algorithm for Polygon-Biasing

The situation for biasing a polygon is sketched in Figure C.1. A polygon is defined as a sequence of points which are given in a counterclockwise order by oriented segments (vectors). A general algorithm to calculate a new "shrinked" or "grown" polygon based on shifting its segments parallel by a certain amount (the bias) has to be applied. This shift is equivalent to the movement of the points of the polygon by a normal vector $\vec{e}_n$ calculated from the normal vectors of the two segments g_{12} and g_{23} times the bias.

This algorithm is only valid for convex polygons. For concave polygons (polygons where at least one point has an internal angle $> 180°$) this algorithm may fail because of point collisions or generation of loops. However, for small bias and suitable geometries the following simple algorithm is applicable.

The line segment $\vec{g}$ can be described by the general parametric line equation:

$$Ax + By + C = 0 \tag{C.1}$$

Furthermore, the normal vector $\vec{e}_n$ is given by

$$\vec{e}_n = A\vec{x}_n + B\vec{y}_n \tag{C.2}$$

and thus the second equation

$$A^2 + B^2 = 1 \tag{C.3}$$

must be fulfilled too. Substituting the start and end coordinates of the vector $\vec{g}_{12}$, $P_1 = (x_1, y_1)$ and $P_2 = (x_2, y_2)$ into (C.1) and (C.3) gives three equations which have to be solved to get the parameters A, B of the line segment.

$$C + Ax_1 + By_1 = 0 \tag{C.4}$$
$$C + Ax_2 + By_2 = 0 \tag{C.5}$$
$$A^2 + B^2 = 1 \tag{C.6}$$
$$\tag{C.7}$$

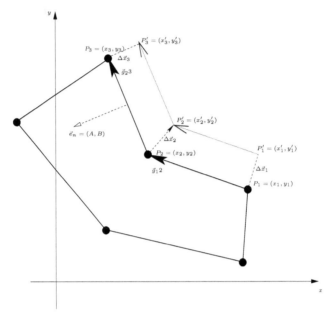

Figure C.1: Geometrical situation when biasing a polygon with a given distance

with the solutions

$$A = \pm \frac{|x_1 - x_2|(y_2 - y_1)}{(x_1 - x_2)\sqrt{(x_1 - x_2)^2 + (y_1 - y_2)^2}} \tag{C.8}$$

$$B = \pm \frac{|x_1 - x_2|}{\sqrt{(x_1 - x_2)^2 + (y_1 - y_2)^2}} \tag{C.9}$$

Per definition the normal vector $\vec{e}_n$ is positive if the line segment $\vec{g}$ is rotated 90° counterclockwise into the normal vector. This means

$$\frac{\vec{g} \times \vec{e}_n}{|\vec{g}|} > 0 \Rightarrow \frac{1}{\sqrt{(x_1 - x_2)^2 + (y_1 - y_2)^2}} \begin{pmatrix} x_2 - x_1 \\ y_2 - y_1 \\ 0 \end{pmatrix} \times \begin{pmatrix} A \\ B \\ 0 \end{pmatrix} > \begin{pmatrix} 0 \\ 0 \\ 0 \end{pmatrix} \tag{C.10}$$

which yields the criteria

$$B \frac{x_2 - x_1}{\sqrt{(x_1 - x_2)^2 + (y_1 - y_2)^2}} - A \frac{y_2 - y_1}{\sqrt{(x_1 - x_2)^2 + (y_1 - y_2)^2}} > 0 \tag{C.11}$$

The relation (C.11) is only fulfilled with the solution

$$A = -\frac{y_2 - y_1}{\sqrt{(x_1 - x_2)^2 + (y_1 - y_2)^2}} \tag{C.12}$$

$$B = \frac{x_2 - x_1}{\sqrt{(x_1 - x_2)^2 + (y_1 - y_2)^2}} \tag{C.13}$$

$$\tag{C.14}$$

of the (C.9),(C.9). To determine the vector by which each point is shifted during biasing (by the distance d), the two line segments $\vec{g}_{12}$ and $\vec{g}_{23}$ attached to the point are moved parallel and thus form the new intersection point P_2'. The normal vectors remain the same, but the parameters C of the general line equations change accordingly to the parallel shift. The initial intersection point $P_2 = (x_2, y_2)$ is moved by the vector $\Delta \vec{x}_2$ to the point $P_2' = (x_2', y_2')$ (see Figure C.1). The parameters C_{12}, C_{23} of the shifted line equations are calculated from their respective line equations and the unchanged line parameters $A_{12}, A_{23}, B_{12}, B_{23}$.

$$C_{12} = -(A_{12}x + B_{12}y + A_{12}^2 d + B_{12}^2) \tag{C.15}$$

$$C_{23} = -(A_{23}x + B_{23}y + A_{23}^2 d + B_{23}^2) \tag{C.16}$$

The new intersection point $P_2' = (x_2', y_2')$ must satisfy both line equations and thus by solving the equations the new intersection point coordinates yield

$$x_2' = \frac{\left(A_{23}^2 B_{12} - \left(A_{12}^2 + B_{12}(B_{12} - B_{23})\right)B_{23}\right)d}{A_{23}B_{12} - A_{12}B_{23}} + x_2 \tag{C.17}$$

$$y_2' = \frac{\left(A_{12}{}^2 A_{23} + A_{23}B_{12}{}^2 - A_{12}\left(A_{23}{}^2 + B_{23}{}^2\right)\right)d}{A_{23}B_{12} - A_{12}B_{23}} + y_2 \tag{C.18}$$

Substituting (C.12),(C.13) in (C.17),(C.18) and using the substitutions $g_{12} = |\vec{g}_{12}|$, $g_{23} = |\vec{g}_{23}|$, $\Delta x_{21} = x_2 - x_1$, $\Delta x_{32} = x_3 - x_2$, $\Delta y_{21} = y_2 - y_1$ and $\Delta y_{32} = y_3 - y_2$ gives finally

$$x_2' = \frac{d\left(-g_{23}\Delta x_{21} + g_{12}\Delta x_{32}\right)}{-x_3\Delta y_{21} - x_1\Delta y_{32} + x_2\Delta y_{31}} + x_2 \tag{C.19}$$

$$y_2' = \frac{d\left(-g_{23}\Delta y_{21} + g_{12}\Delta y_{32}\right)}{-x_3\Delta y_{21} - x_1\Delta y_{32} + x_2\Delta y_{31}} + y_2 \tag{C.20}$$

Appendix D

From Boltzmann Distribution to Drift-Diffusion Current Equations

We consider a steady state situation and, for simplicity, a one-dimenstional geometry [204]. With the use of a relaxation time approximation as in (3.2) the BOLTZMANN equation becomes

$$\frac{eF}{m^*}\frac{\partial f(v,x)}{\partial v} + v\frac{\partial f(v,x)}{\partial x} = \frac{f_{eq}(v,x) - f(v,x)}{\tau} \tag{D.1}$$

Here the relation $m^*\vec{v} = \vec{p} = \hbar\vec{k}$ was used, which is valid for a parabolic energy band. Note that the charge e has to be taken with the proper sign of the particle (positive for holes and negative for electrons). A general definition of current density is given by

$$J(x) = e\int vf(v,x)dv \tag{D.2}$$

where the integral on the right hand side represents the first moment of the distribution function. This definition of current can be related to (D.1). After multiplying both sides by v and integrating over v one gets

$$\frac{eF}{m^*}\int v\frac{\partial f(v,x)}{\partial v}dv + \int v^2\frac{\partial f(v,x)}{\partial x}dv = \frac{1}{\tau}\left[\underbrace{\int vf_{eq}(v,x)dv}_{0} - \int vf(v,x)dv\right] = -\frac{J(x)}{e\tau} \tag{D.3}$$

since the function $vf_{eq}(v,x)$ is odd in v, and its integral is therefore zero. Thus, one has from (D.3)

$$J(x) = -e\frac{e\tau}{m^*}F\int v\frac{\partial f}{\partial v}dv - e\tau\frac{d}{dx}\int v^2 f(v,x)dv \tag{D.4}$$

Integrating by parts yields

$$\int v\frac{\partial f}{\partial v}dv = \underbrace{[vf(v,x)]_{-\infty}^{\infty}}_{0} - \int f(v,x)dv = -n(x) \tag{D.5}$$

147

and

$$\int v^2 f(v,x)dv = n(x)\langle v^2 \rangle \tag{D.6}$$

can be written, where $\langle v^2 \rangle$ is the average of the square of the velocity defined as

$$\langle v^2 \rangle = \frac{1}{n}\int v^2 f(v,x)dv \tag{D.7}$$

Because of the *equipartition theorem*, for a purely one-dimensional treatment, the $-\frac{3}{2}$ exponent in (3.3) may be replaced with $-\frac{1}{2}$, while the appropriate thermal kinetic energy becomes $\frac{k_B T}{2}$ instead of $\frac{3k_B T}{2}$.

The drift-diffusion equations are derived introducing the mobility $\mu = \frac{e\tau}{m^*}$ and replacing $\langle v^2 \rangle$ with its average equilibrium value $\frac{k_B T}{m^*}$, therefore neglecting thermal effects. The diffusion coefficient $D = \frac{\mu k_B T_0}{e}$ (Einstein's relation) is also introduced, and the resulting drift-diffusion current is

$$J_n = qn(x)\mu_n F(x) + qD_n \frac{dn}{dx}$$
$$J_p = qp(x)\mu_p F(x) - qD_p \frac{dp}{dx} \tag{D.8}$$

where q is used to indicate the absolute value of the electronic charge. Although no direct assumptions on the non-equilibrium distribution function $f(v,x)$ was made in the derivation of (D.8), the choice of equilibrium (thermal) velocity means that the drift-diffusion equations are only valid for very small perturbations of the equilibrium state (low fields). The validity of the drift-diffusion equations is *empirically* extended by introducing field-dependent mobility $\mu(F)$ and diffusion coefficient $D(F)$, obtained from experimental measurements.

Appendix E

Diffraction in Far Field Approximation

If the optical wavelength is a linear function of the coordinates of the points in the plane of the aperture and the plane of the projection, one can apply the so called FRAUNHOFER Far Field Approximation. There are many examples for such a case. E.g. the source and the projection planes are far away from the aperture. Another possibility would be the use of lenses. In the following section only the former situation (source and projection planes are in the far field) is considered.

E.1 Linear Approximation

For the following discussion, Figure E.1 shows the situation. The aperture is positioned in the $\tilde{x}\tilde{y}$ plane, which is normal to the z-axis and at the position of point $\tilde{O}$. The point source is

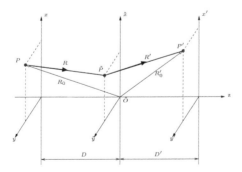

Figure E.1: Standard geometry of theory of diffraction

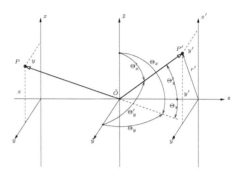

Figure E.2: Viewing angles of source point and projection point as seen from the center of the aperture.

positioned in the xy plane which is parallel to the $\tilde{x}\tilde{y}$ plane in a distance D in the negative z direction. The coordinates of the point source are therefore $(x, y, -D)$. The projection point P' with the coordinates (x', y', D') is in the projection plane $x'y'$, which is parallel to the aperture plane $\tilde{x}\tilde{y}$ also. The positions $\tilde{P}$ in the aperture are defined by the coordinates $(\tilde{x}, \tilde{y}, 0)$. The directions of the incoming and outgoing light waves are given with respect to the center of the aperture plane (see Figure E.2). The directional vectors of the light waves form the angles Θ_x, Θ_y and Θ_z with their respective axes x, y and z. Here we define the functions:

$$\alpha = \cos\Theta_x = \frac{-x}{R_0} = \sin\Theta \tag{E.1}$$

$$\beta = \cos\Theta_y = \frac{-y}{R_0} \tag{E.2}$$

$$\gamma = \cos\Theta_z = \sqrt{1 - \alpha^2 - \beta^2} = \frac{D}{R_0} \tag{E.3}$$

$$\alpha' = \cos\Theta'_x = \frac{-x'}{R'_0} = \sin\Theta' \tag{E.4}$$

$$\beta' = \cos\Theta'_y = \frac{-y'}{R'_0} \tag{E.5}$$

$$\gamma' = \cos\Theta'_z = \sqrt{1 - \alpha'^2 - \beta'^2} = \frac{D'}{R'_0} \tag{E.6}$$

$$\Theta = \frac{\pi}{2} - \Theta_x \tag{E.7}$$

$$\Theta' = \frac{\pi}{2} - \Theta'_x \tag{E.8}$$

If one defines the transmission function:

$$\tilde{\tau}(\tilde{x}, \tilde{y}) = \begin{cases} 1 & : & (\tilde{x}, \tilde{y}) \in \tilde{\sigma}_0 \\ 0 & : & (\tilde{x}, \tilde{y}) \in \tilde{\sigma}_c \end{cases} \tag{E.9}$$

with $\sigma = \sigma_0 \vee \sigma_c$ and $\tilde\sigma_0$ and $\tilde\sigma_c$ being the aperture area which are transmitting and not transmitting light respectively. The resulting electric field on the projection side after diffraction at an aperture given by the transmission function $\tilde\tau$ yields

$$E' = C \iint\limits_{\tilde\sigma} \tilde\tau \tilde E \frac{e^{-ikR'}}{R'} \, d\tilde\sigma \tag{E.10}$$

taking the solution of a spherical wave for diffraction of a planar wave at an infinite small aperture opening

$$\tilde E = \frac{A}{R} e^{i(\omega t - kR)} \tag{E.11}$$

with A as the Amplitude of the incident planar wave. Equation (E.10) yields

$$E' = C A e^{i\omega t} \int\limits_{\tilde\sigma} \tau \frac{e^{-ik(R+R')}}{RR'} \, d\tilde\sigma \tag{E.12}$$

The lengths R and R' are given by

$$\begin{aligned} R &= \sqrt{(\tilde x - x)^2 + (\tilde y - y)^2 + D^2} \\ R' &= \sqrt{(x' - \tilde x)^2 + (y' - \tilde y)^2 + D'^2} \end{aligned} \tag{E.13}$$

For the far field approximation R and R' are linear functions of $\tilde x$ and $\tilde y$. Therefore we calculate the series expansion of R and R' around R_0 and R'_0, with R_0 and R'_0 given by

$$\begin{aligned} R_0 &= \sqrt{x^2 + y^2 + D^2} \\ R'_0 &= \sqrt{x'^2 + y'^2 + D'^2} \end{aligned} \tag{E.14}$$

These distances are no functions of $\tilde x$ and $\tilde y$. Using (E.14) in (E.13) yields

$$R = \sqrt{R_0^2 - 2(x\tilde x - y\tilde y) + \tilde x^2 + \tilde y^2} = R_0 \sqrt{1 - \frac{2}{R_0^2}(x\tilde x + y\tilde y) + \frac{\tilde x^2 + \tilde y^2}{R_0^2}} \tag{E.15}$$

assuming that the second and third term is much smaller than the first one, the squareroot in (E.15) can be written as

$$R = R_0 \sqrt{1 + \varepsilon} \simeq R_0 (1 + \frac{1}{2}\varepsilon - \frac{1}{8}\varepsilon^2) \tag{E.16}$$

and

$$\varepsilon = -\frac{2}{R_0^2}(x\tilde x + y\tilde y) + \frac{\tilde x^2 + \tilde y^2}{R_0^2} \tag{E.17}$$

Finally R is calculated by the series expansion to the second order in $\tilde x$ and $\tilde y$

$$R \simeq R_0 (1 - \frac{x\tilde x + y\tilde y}{R_0^2} + \frac{\tilde x^2 + \tilde y^2}{2R_0^2} - \frac{(x\tilde x + y\tilde y)^2}{2R_0^4}) \tag{E.18}$$

This result is only valid if $\tilde{x} \ll R_0$ and $\tilde{y} \ll R_0$. This holds especially for the far field where R_0 is big in relation to the aperture size

$$
\begin{aligned}
|\tilde{x}| &\ll \sqrt{R_0 \lambda} \\
|\tilde{y}| &\ll \sqrt{R_0 \lambda}
\end{aligned}
\tag{E.19}
$$

Therefore the third term in (E.18) is negligible and this gives

$$
\begin{aligned}
R &\simeq R_0 - (\frac{x}{R_0})\tilde{x} - (\frac{y}{R_0})\tilde{y} \\
R' &\simeq R'_0 - (\frac{x'}{R'_0})\tilde{x} - (\frac{y'}{R'_0})\tilde{y}
\end{aligned}
\tag{E.20}
$$

Using the approximation

$$
(RR')^{-1} \simeq (R_0 R'_0)^{-1}
\tag{E.21}
$$

which is valid for the far field, and the transformation of the coordinates

$$
\begin{aligned}
u &\equiv -(\frac{x}{R_0} + \frac{x'}{R'_0})\frac{1}{\lambda} = \frac{(\alpha - \alpha')}{\lambda} = \frac{\cos\theta_x - \cos\theta'_x}{\lambda} \\
v &\equiv -(\frac{y}{R_0} + \frac{y'}{R'_0})\frac{1}{\lambda} = \frac{(\beta - \beta')}{\lambda} = \frac{\cos\theta_y - \cos\theta'_y}{\lambda}
\end{aligned}
\tag{E.22}
$$

the total optical path can be written as

$$
R + R' = (R_0 + R'_0) + (u\tilde{x} + v\tilde{y})\lambda
\tag{E.23}
$$

substituting this optical path in the phase factor of the diffraction integral in (E.12) yields

$$
e^{i\omega t} e^{-ik(R+R')} = e^{i\phi_0} e^{-i2\pi(u\tilde{x}+v\tilde{y})}
\tag{E.24}
$$

with

$$
\phi_0 = \omega t - k(R_0 + R'_0)
\tag{E.25}
$$

ϕ_0 is not a function of $\tilde{x}$ and $\tilde{y}$. The coordinates (x, y) of the source and (x', y') of the projection point are now included in (u, v). Finally the electrical field at the projection point P' emanating from the source point P and diffracted at the aperture with the transmission function $\tilde{\tau}(\tilde{x}, \tilde{y})$ can be given as

$$
E'(u, v) = \frac{CAe^{i\phi_0}}{R_0 R'_0} \int\!\!\!\int_{-\infty}^{+\infty} \tilde{\tau}(\tilde{x}, \tilde{y}) e^{i2\pi(u\tilde{x}+v\tilde{y})} \, d\tilde{x} \, d\tilde{y}
\tag{E.26}
$$

The integral of (E.26) is exactly the FOURIER transform of the transmission function $\tilde{\tau}(\tilde{x}, \tilde{y})$

$$
T(u, v) \equiv \int\!\!\!\int_{-\infty}^{+\infty} \tilde{\tau}(\tilde{x}, \tilde{y}) e^{i2\pi(u\tilde{x}+v\tilde{y})} \, d\tilde{x} \, d\tilde{y}
\tag{E.27}
$$

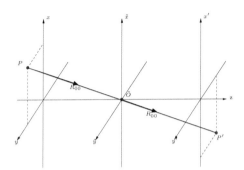

Figure E.3: Direct light propagation through aperture

This important result implies, that under the given assumptions, *the electrical field distribution after a diffracting aperture is proportional to the* FOURIER *transform of the transmission function of the aperture.*

If as shown in Figure E.3 the points $P, \tilde{O}$ and P' are on a line, then $(u, v) = (0, 0)$ and (E.26) together with (E.27) reduces to

$$E'(0,0) = \frac{C A e^{i\phi_{00}}}{R_{00} R'_{00}} T(0,0) \tag{E.28}$$

With (E.28), (E.26) can be normalized to $E'(0,0)$

$$E'(u,v) = E'(0,0) \left[\frac{R_{00} R'_{00}}{R_0 R'_0} \right] \frac{T(u,v)}{T(0,0)} e^{i(\phi_0 - \phi_{00})} \tag{E.29}$$

E.2 Circular Aperture

Figure E.4 shows the aperture plane with the coordinates $\tilde{r}$ and $\tilde{\phi}$. The transmission function of the aperture is

$$\tilde{\tau}(\tilde{r}, \tilde{\phi}) = \begin{cases} 1 & : & \tilde{r} < \tilde{r}_0 \\ 0 & : & \tilde{r} \geq \tilde{r}_0 \end{cases} \tag{E.30}$$

and

$$\begin{aligned} \tilde{x} &= \tilde{r} \cos \tilde{\phi} \\ \tilde{y} &= \tilde{r} \sin \tilde{\phi} \end{aligned} \tag{E.31}$$

Using polar coordinates for the transformed coordinates u and v defined in the previous section

$$\begin{aligned} u &= \rho \cos \phi' \\ v &= \rho \sin \phi' \end{aligned} \tag{E.32}$$

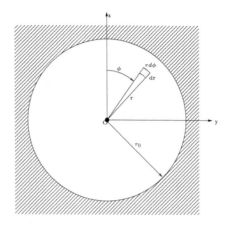

Figure E.4: Coordinate system in a circular aperture

gives

$$\rho = \sqrt{u^2 + v^2} = \frac{1}{\lambda}\sqrt{\left(\frac{x}{R_0} + \frac{x'}{R_0'}\right)^2 + \left(\frac{y}{R_0} + \frac{y'}{R_0'}\right)^2} \tag{E.33}$$

We are now reviewing the case with the source being on the optical axis of the aperture ($x = y = 0$) which yields

$$\rho = \frac{1}{\lambda}\frac{r'}{R_0'} = \frac{\sin\theta_z'}{\lambda} \tag{E.34}$$

where

$$r' = \sqrt{x'^2 + y'^2} \tag{E.35}$$

is the radial distance of the projection point from the z-axis in the projection plane. θ_z' is the angle of the projection point seen from the aperture (refer to Figure E.2). The FOURIER transform of $\tilde{\tau}$ can be calculated from (E.27). The differential area element $d\tilde{\sigma}$ is in polar coordinates

$$d\tilde{\sigma} = \tilde{r}\,d\tilde{\phi}\,d\tilde{r}$$

as depicted by Figure E.4 The exponential factor in the integral of (E.27) in the above defined coordinate system gives

$$
\begin{aligned}
\exp[-i2\pi(\tilde{u}x + \tilde{v}y)] &= \exp[-i2\pi\rho\tilde{r}(\cos\phi'\cos\tilde{\phi} + \sin\phi'\sin\tilde{\phi})] \\
&= \exp[-i2\pi\rho\tilde{r}\cos(\tilde{\phi} - \phi')]
\end{aligned}
$$

Therefore (E.27) is in polar coordinates

$$T(\rho, \phi') \cong \int_0^{\tilde{r}_0}\tilde{r}\int_0^{2\pi}\exp[-i2\pi\tilde{r}\cos(\tilde{\phi} - \phi')]\,d\tilde{\phi}\,d\tilde{r} \tag{E.36}$$

The inner integrand of this solution is the well known BESSEL function of zeroth order and is defined as

$$J_0(w) = \frac{1}{2\pi} \int\limits_0^{2\pi} \exp[-iw\cos(\tilde{\phi} - \phi')]\, d\tilde{\phi} \tag{E.37}$$

by using this definition with (E.27) the FOURIER transform of a circular aperture gives

$$T(\rho) = 2\pi \int\limits_0^{\tilde{r}_0} \tilde{r} J_0(2\pi\rho\tilde{r})\, d\tilde{r} \tag{E.38}$$

which is independent of ρ' as a consequence of the rotational symmetry of the aperture. Applying a coordinate transformation $w' = 2\pi\rho\tilde{r}$ to the integral of (E.38) yields

$$T(\rho) = \frac{1}{2\pi\rho^2} \int\limits_0^{2\pi\rho\tilde{r}_0} w' J_0(w')\, dw' \tag{E.39}$$

To solve this integral one can use a relation between BESSEL functions of different order

$$w\frac{dJ_n}{dw} + nJ_n = wJ_{n-1} \tag{E.40}$$

integrating (E.40) with $n = 1$ gives

$$w\frac{J_1}{dw} + J_1 = wJ_0$$
$$\frac{d(wJ_1)}{dw} = wJ_0$$
$$wJ_1(w) = \int\limits_0^w w' J_0(w')\, dw' \tag{E.41}$$

Using this result in (E.39) gives finally

$$T(\rho) = \frac{1}{2\pi\rho^2} 2\pi\rho\tilde{r}_0 J_1(2\pi\rho\tilde{r}_0) \tag{E.42}$$

or

$$T(\rho) = 2\pi\tilde{r}_0{}^2 \frac{J_1(2\pi\rho\tilde{r}_0)}{2\pi\rho\tilde{r}_0} \tag{E.43}$$

In the center of the diffraction pattern with $\rho = 0$ the properties of the BESSEL function give

$$\lim_{w\to 0}\left(\frac{J_1(w)}{w}\right) = \frac{1}{2}$$

yielding

$$T(0) = \pi\tilde{r}_0{}^2 \tag{E.44}$$

Taking (E.29) the electrical field of the diffraction pattern behind a circular aperture is finally

$$E'(\rho) = E'(0) \left(\frac{2J_1(2\pi\rho\tilde{r}_0)}{2\pi\rho\tilde{r}_0} \right) \left(\frac{D'}{R'_0} \right) e^{i(\phi_0 - \phi_{00})} \tag{E.45}$$

The average optical intensity is proportional to the square of the absolute electrical field. Therefore the intensity for diffraction behind a circular aperture is

$$I'(\rho) = I'(0) \left(\frac{2J_1(2\pi\rho\tilde{r}_0)}{2\pi\rho\tilde{r}_0} \right)^2 \left(\frac{D'}{R'_0} \right)^2 \tag{E.46}$$

with

$$\rho = \frac{\sin\theta'_z}{\lambda} = \frac{r'}{R'_0\lambda} \tag{E.47}$$

By using the (E.14) and (E.35) in (E.46) one obtains

$$I'(\rho) = I'(0) \left(\frac{2J_1(2\pi\rho\tilde{r}_0)}{2\pi\rho\tilde{r}_0} \right)^2 \frac{1}{1 + \left(\frac{r'}{D'} \right)^2} \tag{E.48}$$

Substituting r' from (E.47) into above equation yields

$$I'(\rho) = I'(0) \left(\frac{2J_1(2\pi\rho\tilde{r}_0)}{2\pi\rho\tilde{r}_0} \right)^2 \left(1 - \rho^2\lambda^2 \right) \tag{E.49}$$

With the substitution

$$x = 2\pi\rho\tilde{r}_0 = \frac{2\pi\tilde{r}_0 r'}{R'_0\lambda} \tag{E.50}$$

(which is dimensionless) the intensity is finally

$$I'(\rho) = I'(0) \left(\frac{2J_1(x)}{x} \right)^2 \left(1 - x^2 \left[\frac{\lambda}{2\pi\tilde{r}_0} \right]^2 \right) \tag{E.51}$$

The (for most cases valid) assumption that the aperture radius is much bigger than the wavelength ($\lambda \ll \tilde{r}_0$) the intensity is

$$I'(\rho) = I'(0) \left(\frac{2J_1(x)}{x} \right)^2 \tag{E.52}$$

Which is the well known intensity distribution for a circular aperture given in many textbooks. The variable x reduces to

$$x = \frac{2\pi\tilde{r}_0 r'}{D'\lambda} \tag{E.53}$$

for the assumption mentioned above.

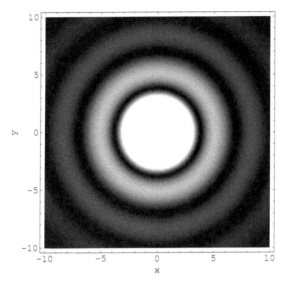

Figure E.5: AIRY-disk with rings

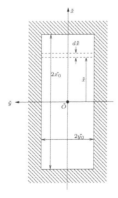

Figure E.6: Coordinate system in a rectangular aperture

E.3   Rectangular Aperture

Figure E.6 shows the aperture plane with the coordinates $\tilde{x}$ and $\tilde{y}$. The transmission function is

$$\tilde{\tau}(\tilde{x}, \tilde{y}) = \begin{cases} 1 & : \quad |\tilde{x}| < \tilde{y}_0 \wedge |\tilde{y}| < \tilde{y}_0 \\ 0 & : \quad otherwise \end{cases} \tag{E.54}$$

The area of the aperture is given by $4\tilde{x}_0\tilde{y}_0$. The FOURIER transform can be calculated from (E.27) and splitted into two integrals

$$T(u, v) = T(u)T(v) = \int\limits_{-\tilde{x}_0}^{\tilde{x}_0} e^{-i2\pi u\tilde{x}} \, d\tilde{x} \int\limits_{-\tilde{y}_0}^{\tilde{y}_0} e^{-i2\pi v\tilde{y}} \, d\tilde{y} \tag{E.55}$$

One stripe in Figure E.6 is thereby given by

$$e^{-i2\pi u\tilde{x}} \, d\tilde{x} \int\limits_{-\tilde{y}_0}^{\tilde{y}_0} e^{-i2\pi v\tilde{y}} \, d\tilde{y} \tag{E.56}$$

Integration of (E.55) is simple and straightforward

$$T(v) = \frac{1}{-i2\pi v} \left(e^{-i2\pi v\tilde{y}_0} - e^{+i2\pi v\tilde{y}_0} \right) = \frac{\sin(2\pi v\tilde{y}_0)}{\pi v} = 2\tilde{y}_0 \frac{\sin(2\pi v\tilde{y}_0)}{2\pi v\tilde{y}_0} \tag{E.57}$$

The rightmost term can be defined as a new function

$$\operatorname{sinc}(w) \equiv \frac{\sin(w)}{w} \tag{E.58}$$

With this substitution (E.57) yields

$$T(v) = 2\tilde{y}_0 \operatorname{sinc}(2\pi v\tilde{y}_0) \tag{E.59}$$

and

$$T(u) = 2\tilde{x}_0 \operatorname{sinc}(2\pi u\tilde{x}_0) \tag{E.60}$$

in an analogous way. Therefore in Point P' the electric field is according to (E.29)

$$\begin{aligned} E'(u, v) &= E'(0, 0)\frac{2x_0 \operatorname{sinc}(2\pi u\tilde{x}_0)2y_0 \operatorname{sinc}(2\pi\tilde{y}_0)}{4\tilde{x}_0\tilde{y}_0} \left[\frac{R_{00}R'_{00}}{R_0 R'_0} \right] e^{i(\phi_0 - \phi_{00})} \\ &= E'(0, 0) \operatorname{sinc}(2\pi u\tilde{x}_0) \operatorname{sinc}(2\pi\tilde{y}_0) \left[\frac{R_{00}R'_{00}}{R_0 R'_0} \right] e^{i(\phi_0 - \phi_{00})} \end{aligned} \tag{E.61}$$

and the intensity as the square of the electrical field is then

$$I'(u, v) = I'(0, 0) \operatorname{sinc}^2(2\pi u\tilde{x}_0) \operatorname{sinc}^2(2\pi v\tilde{y}_0) \left[\frac{R_{00}R'_{00}}{R_0 R'_0} \right]^2 \tag{E.62}$$

For the special case of the source being located on the z-axis, the coordinates x, y are zero and the coordinates x', y' are the following functions of u, v

$$x' = -R_0'\lambda u \quad \text{and} \quad y' = -R_0'\lambda v \quad . \tag{E.63}$$

For R_{00} and R_{00}' (the direct beam) the coordinates are $u = v = 0$ and according to (E.63) $x' = y' = 0$. The direct beam is therefore in the z-axis and R_{00}, R_{00}' can be written as

$$\begin{aligned}
R_{00} &\equiv D \\
R_{00}' &\equiv D'
\end{aligned} \tag{E.64}$$

Together with (E.14) and (E.64) the square of the fraction in (E.62) gives

$$\begin{aligned}
\left(\frac{R_{00}R_{00}'}{R_0 R_0'}\right)^2 &= \frac{D^2 D'^2}{(x^2 + y^2 + D^2)(x'^2 + y'^2 + D'^2)} \\
&= \frac{1}{(\frac{x^2+y^2}{D^2} + 1)(\frac{x'^2+y'^2}{D'^2} + 1)} \overset{(x=y=0)}{\Rightarrow} \\
\Rightarrow \left(\frac{R_{00}R_{00}'}{R_0 R_0'}\right)^2 &= \frac{1}{\frac{x'^2+y'^2}{D'^2} + 1}
\end{aligned} \tag{E.65}$$

Therefore this fraction can be set to unity if $|x'| \ll D' \wedge |y'| \ll D'$. This assumption yields finally for the intensity behind a rectangular aperture

$$\begin{aligned}
I'(x', y') &= I'(0,0)\,\text{sinc}^2\left(\frac{-2\pi x'\tilde{x}_0}{R_0'\lambda}\right)\text{sinc}^2\left(\frac{-2\pi y'\tilde{y}_0}{R_0'\lambda}\right) \\
&\cong I'(0,0)\,\text{sinc}^2\left(\frac{2\pi x'\tilde{x}_0}{D'\lambda}\right)\text{sinc}^2\left(\frac{2\pi y'\tilde{y}_0}{D'\lambda}\right)
\end{aligned} \tag{E.66}$$

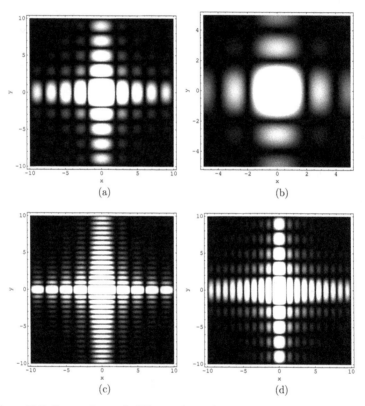

Figure E.7: Comparison of different intensity distributions after diffraction at a rectangular aperture (a) square aperture (b) detail of square aperture (c) rectangular aperture with $\tilde{x}_0 = 2\tilde{y}_0$ (d) rectangular aperture with $\tilde{y}_0 = 2\tilde{x}_0$

Bibliography

[1] International Semiconductor Industry Association, "International Technology Roadmap for Semiconductors 2005 - Modeling & Simulation," 2005.
URL: http://www.itrs.net/Common/2005ITRS/Modeling2005.pdf.

[2] G. Moore, "Cramming More Components onto Integrated Circuits," *Proc.IEEE*, vol. 86, no. 1, pp. 82–85, 1998.

[3] IC Knowledge LLC, *Microprocessor Trends*.
URL: http://www.icknowledge.com/trends/uproc.html.

[4] International Semiconductor Industry Association, "International Technology Roadmap for Semiconductors 2004 Update - Overall Roadmap Technology Characteristics," 2004.
URL: http://www.itrs.net/Common/2004Update/2004_000_ORTC.pdf.

[5] R. Dutton and Y. Yu, *Technology CAD - Computer Simulation of IC Processes and Device*, vol. 1. Norwell: Kluwer, 1993.

[6] S. Cazzani, "A SPICE Simulator for Complex Circuits," *Elettronica-Oggi*, no. 164, pp. 112–114, 1993.

[7] B. Hennion, Y. Paradis, and P. Senn, "ELDO: A General Purpose Third Generation Circuit Simulator based on the OSR Method," in *Proceedings of the European Conference on Circuits Theory and Design - ECCTD 87*, (North-Holland, Amsterdam, Netherlands), pp. 113–120, 1987.

[8] R. A. Gough and P. C. Marston, "Signal Processing Simulated by a SPECTRE," *MSN-Microwave-Systems-News*, vol. 13, no. 2, pp. 57–74, 1983.

[9] Synopsys, Inc., Mountain View, California, *TCAD Software*.
URL: http://www.synopsys.com.

[10] Silvaco, Int., Santa Clara, California, *TCAD Software*.
URL: http://www.silvaco.com.

[11] R. Strasser, *Rigorous TCAD Investigations on Semiconductor Fabrication Technology*. Dissertation, Technische Universität Wien, 1999. URL: http://www.iue.tuwien.ac.at/phd/strasser.

[12] D. Sohl and P. Kumar, "Fluctuation smoothing scheduling policies for multiple process flow fabrication plants," in *Seventeenth IEEE/CPMT International Electronics Manufacturing Technology Symposium*, (New York, NY, USA), pp. 190–198, IEEE, 1997.

[13] M. Meghelli, M. Bouche, and A. Konczykowska, "Very High Speed Integrated Circuits: Design Methodology and Applications for Optical Communications," in *ECCTD '97. Proceedings of the 1997 European Conference on Circuit Theory and Design*, vol. 3, (Budapest, Hungary), pp. 1366–1370, Tech. Univ. Budapest, 1997.

[14] S. Das, A. Chandrakasan, and R. Reif, "Three-Dimensional Integrated Circuits: Performance, Design Methodology, and CAD Tools," in *Proceedings IEEE Computer Society Annual Symposium on VLSI. New Trends and Technologies for VLSI Systems Design*, (Los Alamitos, CA, USA), pp. 13–18, IEEE Comput. Soc., 2003.

[15] R. Beniseviciute, "The Aspects of Non-Standard CMOS Integrated Circuits Layout Design," *Elektronika i Elektrotechnika*, vol. 2, pp. 43–6, 2005.

[16] R. Prasad and I. Koren, "Constructive Floorplanning with a Yield Objective," in *Proceedings of the IEEE 2001. 2nd International Symposium on Quality Electronic Design*, (Los Alamos, CA, USA), pp. 261–266, IEEE Comput. Soc., 2001.

[17] M. Bohr, "Scaling of High Performance Interconnects," in *Advanced Metallization and Interconnect Systems for ULSI Applications in 1996*, (Pittsburgh, PA, USA), pp. 3–10, Material Research Society, 1997.

[18] G. A. Burns and J. A. Schoeffel, "Performance Evaluation of the ATEQ CORE-2000 Scanning Laser Reticle Writer," in *Proceedings-of-the-SPIE–The-International-Society-for-Optical-Engineering*, vol. 772, pp. 269–276, 1987.

[19] A. Fujii, K. Mizuno, T. Nakahara, S. Asai, Y. Kadowaki, H. Shimada, H. Touda, K. Iizumi, H. Takahasi, K. Oonuki, T. Kawahara, K. Kawasaki, K. Nagata, and H. Satoh, "Advanced E-Beam Reticle Writing System for Next Generation Reticle Fabrication," in *Proceedings-of-the-SPIE–The-International-Society-for-Optical-Engineering*, vol. 4409, pp. 258–269, 2001.

[20] H. Geiler, H. Karge, M. Wagner, A. Ehlert, E. Daub, and K. Messmann, "Detection and Analysis of Crystal Defects in Silicon by Scanning Infrared Depolarization and Photoluminescence Heterodyne Techniques," *Materials Science & Engineering B Solid State Materials for Advanced Technology*, vol. B, pp. 46–50, April 2002.

[21] M. Swaminathan, "Challenges in ECAD Tool Integration for Digital and RF Packages," in *Proceedings of the Pacific Rim/ASME International Intersociety Electronic and Photonic Packaging Conference - INTERpack '97*, vol. 1, (New York, NY, USA), p. 883, 1997.

[22] J. Ladvanszky, "On Package Modeling," in *Proceedings of the 24th European Microwave Conference*, vol. 2, (Nexus Business Community, Swanley, UK), pp. 1638–1643, 1994.

[23] F. Nusseibeh, S. Ahderom, D. Parent, M. Gokhale, F. Jain, P. Robinson, and D. Mahulikar, "Three-Dimensional Modeling of MQUAD Packages at 100 MHz," in *Proceedings of the International Electronics Packaging Conference*, (Wheaton, IL, USA), pp. 385–389, 1995.

[24] T. Roessler and J. Thiele, "Geometrical Analysis of Product Layout as a Powerful Tool for DFM," in *Proceedings of the SPIE The International Society for Optical Engineering*, vol. 1, pp. 150–160, SPIE-Int. Soc. Opt. Eng, 2005.

[25] J. Torres and C. Berglund, "Integrated Circuit DFM Framework for Deep Sub-Wavelength Processes," in *Proceedings of the SPIE The International Society for Optical Engineering*, vol. 1, pp. 46–57, SPIE-Int. Soc. Opt. Eng, 2005.

[26] Mentor Graphics Corp., Wilsonville, Oregon, *Mentor Graphics Homepage*. URL: http://www.mentor.com.

[27] Cadence Design Systems Inc., San Jose, California, *Cadence Homepage*. URL: http://www.cadence.com.

[28] Agilent Technologies Inc., Palo Alto, California, *Agilent EEsof EDA Homepage*. URL: http://eesof.tm.agilent.com.

[29] Ming-Dou-Ker, Tung-Yang-Chen, and Chung-Yu-Wu, "Design and Analysis of On-Chip ESD Protection Circuit with very Low Input Capacitance for High-Precision Analog Applications," *Analog Integrated Circuits and Signal Processing*, vol. 32, pp. 257–278, Sept. 2002.

[30] J. Bruning, "Optical Lithography-Thirty Years and Three Orders of Magnitude," in *Proceedings SPIE's 22nd International Symposium on Microlithography*, pp. 14–27, 1997.

[31] H. Levinson and W. Arnold, "Optical Lithography," in *Handbook of Microlithography, Micromachining and Microfabrication* (P. Rai-Choudhury, ed.), vol. 1, pp. 13–127, New York: SPIE, 1997.

[32] B. Smith, "Microlithography," in *Microlithography: Science and Technology* (J. Sheats and B. Smith, eds.), vol. 1, pp. 1–106, New York: Marcel Dekker, 1998.

[33] J. Bruning, "Optical Imaging for Microfabrication," *Journal of Vacuum Science & Technology B*, vol. 17, no. 5, pp. 1147–1155, 1980.

[34] P. Rumsby, E. Harvey, and D. Thomas, "Laser Microprojection for Micromechanical Device Fabrication," in *Proceedings of the SPIE*, vol. 2921, pp. 684–692, Dec 1997.

[35] W. Trybula and D. Dance, "Cost of Mask Fabrication," in *Proceedings of the SPIE The International Society for Optical Engineering*, pp. 211–215, SPIE-Int. Soc. Opt. Eng, 1997.

[36] M. Sturans, J. Hartley, H. Pfeiffer, R. Dhaliwal, T. Groves, J. Pavik, R. Quickie, C. Clement, and G. Dick, "EL5: One Tool for Advanced X-Ray and Chrome on Glass Mask Making," *Journal of Vacuum Science & Technology B Microelectronics and Nanometer Structures*, vol. 16, pp. 3164–3167, Nov.-Dec. 1998.

[37] Hao-Hsing-Lu, R. Hwang, V. Lee, J. Chen, T. Laidig, K. Wampler, and R. Caldwell, "Making High Performance Scattering Bar OPC Masks with Vector Scan, Variable-Shaped E-Beam and Raster Scan Laser Mask Writers," in *Proceedings of the SPIE The International Society for Optical Engineering*, pp. 368–382, SPIE-Int. Soc. Opt. Eng, 1998.

[38] G. Hughes, D. Kamaruddin, K. Nakagawa, S. MacDonald, W. Wilkinson, C. West, and K. Park, "Study of OPC for AAPSM Reticles Using Various Mask Fabrication Techniques," in *Proceedings of the SPIE The International Society for Optical Engineering*, vol. 5377, pp. 204–211, SPIE-Int. Soc. Opt. Eng, 2004.

[39] S. Tuan, G. Gottlib, and A. Rosenbusch, "The Process of Manufacturing and Inspection of High-End (Ternary) Tritone EAPSM Reticles for 0.13 μm Design Rule Generation," in *Proceedings of the SPIE The International Society for Optical Engineering*, vol. 4754, pp. 422–427, SPIE-Int. Soc. Opt. Eng, 2002.

[40] G. Airy, "On the Diffraction of an Object-Glass with Circular Aperture," *Transactions of the Cambridge Philosophical Society 1834*, vol. 5, no. 3, pp. 283–291, 1834.

[41] J. Strutt, "Investigations in Optics, with Special Reference to the Spectroscope," *Philosophical Magazine*, vol. 8, no. 5, p. 403, 1879.

[42] H. Hopkins, "Numerical Evaluation of the Frequency Response of Optical Systems," *Proceedings of the Physical Society B*, no. 70, pp. 1002–1005, 1957.

[43] S. Kyoh, S. Tanaka, S. Inoue, I. Higashikawa, I. Mori, K. Okumura, N. Irie, K. Muramatsu, Y. Ishii, N. Magome, and T. Umatate, "Management of Pattern Generation System Based on i-Line Stepper," in *Proceedings of the SPIE The International Society for Optical Engineering*, vol. 4186, pp. 494–502, PIE-Int. Soc. Opt. Eng, 2001.

[44] H. Kirchauer, *Photolithography Simulation*. Dissertation, Technische Universität Wien, 1998.
URL: http://www.iue.tuwien.ac.at/phd/kirchauer.

[45] G. Hobler, *Physikalische Modellierung der Ionenimplantation in Silizium*. Habilitationsschrift, Technische Universität Wien, 1995.

[46] H. Stippel, *Simulation der Ionen-Implantation*. Dissertation, Technische Universität Wien, 1993.
URL: http://www.iue.tuwien.ac.at/phd/stippel.

[47] G. Hobler, *Simulation der Ionenimplantation in ein-, zwei- und dreidimensionalen Strukturen*. Dissertation, Technische Universität Wien, 1988.
URL: http://www.iue.tuwien.ac.at/phd/hobler.

[48] A. Hössinger, *Simulation of Ion Implantation for ULSI Technology*. Dissertation, Technische Universität Wien, 2000.
URL: http://www.iue.tuwien.ac.at/phd/hoessinger.

[49] M. Heinrich, *Simulation der Diffusion in Silizium*. Dissertation, Technische Universität Wien, 1991.
URL: http://www.iue.tuwien.ac.at/phd/heinrich.

[50] M. Radi, *Three-Dimensional Simulation of Thermal Oxidation*. Dissertation, Technische Universität Wien, 1998.
URL: http://www.iue.tuwien.ac.at/phd/radi.

[51] B. Lojek, "History of Semiconductors Diffusion Engineering," in *10th IEEE International Conference on Advanced Thermal Processing of Semiconductors RTP 2002*, (Piscataway, NJ, USA), pp. 209–241, IEEE, 2002.

[52] E. Strasser, *Simulation von Topographieprozessen in der Halbleiterfertigung*. Dissertation, Technische Universität Wien, 1994.
URL: http://www.iue.tuwien.ac.at/phd/strasser.

[53] W. Pyka, *Feature Scale Modeling for Etching and Deposition Processes in Semiconductor Manufacturing*. Dissertation, Technische Universität Wien, 2000.
URL: http://www.iue.tuwien.ac.at/phd/pyka.

[54] A. Belov and H. Jager, "Relaxation Kinetics in Amorphous Carbon Films: An Insight from Atomic Scale Simulation," *Thin Solid Films*, vol. 482, pp. 74–78, June 2005.

[55] T. Kakinaga, A. Hatai, O. Tabata, and Y. Isono, "Silicon Anisotropic Wet Etching Simulation Using Molecular Dynamics," in *TRANSDUCERS '05. The 13th International Conference on Solid State Sensors, Actuators and Microsystems*, vol. 1, (Piscataway, NJ, USA), pp. 816–819, IEEE, 2005.

[56] C. Cardinaud, M. Peignon, and P. Tessier, "Plasma Etching: Principles, Mechanisms, Application to Micro- and Nano-Technologies," *Applied Surface Science*, vol. 164, pp. 72–83, Sept 200.

[57] B. Choo, T. Riley, B. Schulz, and B. Singh, "Automated Process Control Monitor for $0.18\,\mu m$ Technology and Beyond," in *Proceedings of the SPIE The International Society for Optical Engineering*, vol. 3998, pp. 218–226, SPIE-Int. Soc. Opt. Eng, 2000.

[58] Meng-Chiou-Wu and Rung-Bin-Lin, "A Comparative Study on Dicing of Multiple Project Wafers," in *Proceedings. IEEE Computer Society Annual Symposium on VLSI* (A. Smailagic and N. Ranganathan, eds.), (Los Alamitos, CA, USA), pp. 314–315, IEEE Comput. Soc, 2005.

[59] A. Kerber and M. Kerber, "Fast Wafer Level Data Acquisition for Reliability Characterization of sub-100 nm CMOS Technologies," in *2004 IEEE International Integrated Reliability Workshop Final Report*, (Piscataway, NJ, USA), pp. 41–45, IEEE, 2004.

[60] Hung-Chin-Lin and Gwo-Ji-Sheen, "Practical Implementation of the Capability Index $C_p k$ Based on the Control Chart Data," *Quality Engineering*, vol. 17, pp. 371–390, July-Sept. 2005.

[61] S. Tohnai, M. Yamazaki, and J. Suzuki, "Realtime Statistical Process Control by QWACS," *NEC Technical Journal*, vol. 51, pp. 208–211, March 1998.

[62] R. Mason, R. Gunst, and J. Hess, *Statistical Design and Analysis of Experiments*, ch. 3, p. 79. Wiley Series in Probability and Statistics, Hoboken, New Jersey: John Wiley & Sons Inc., 2nd ed., 2003.

[63] W. Shewhart, *Economic Control of Quality of Manufactured Product.* Milwaukee, Wisconsin: American Society for Quality Control, 1931, 1980.

[64] L. Chang, "Systematic Methodology with DFT Rules Reduces Fault-Coverage Analysis," *Integrated System Design,* vol. 14, pp. 42–53, Aug. 2001.

[65] R. Anzivino, "Design for Fabrication, Assembly, and Test," in *PCB Design Conference. Proceedings of the Third Annual PCB Design Conference,* vol. 1, (San Francisco, CA, USA), pp. 11–18, Miller Freeman, 1994.

[66] G. Trucco, G. Boselli, and V. Liberalli, "A Study of Crosstalk Through Bonding and Package Parasitics in CMOS Mixed Analog-Digital Circuits," in *Integrated Circuit and System Design. Power and Timing Modeling, Optimization and Simulation. 14th International Workshop, PATMOS 2004. Proceedings Lecture Notes in Comput. Sci.,* vol. 3254, (Berlin, Germany), pp. 138–147, Springer Verlag, 2004.

[67] H. DeGersem, H. Sande, and K. Hameyer, "Strong Coupled Multi-Harmonic Finite Element Simulation Package," *COMPEL The International Journal for Computation and Mathematics in Electrical and Electronic Engineering,* vol. 20, no. 2, pp. 535–546, 2001.

[68] R. Krondorfer, Y. Kim, C. Gustafson, and T. Lommasson, "Finite Element Simulation of Package Stress in Transfer Molded MEMS Pressure Sensors," *Microelectronics Reliability,* vol. 44, no. 12, pp. 1995–2002, 2004.

[69] R. Tummala, E. Rymaszewski, and A. Klopfenstein, *Microelectronics Packaging Handbook,* vol. I - Technology Drivers. New York, NY: Chapman and Hall, 1997.

[70] R. Dutton, "TCAD Tools at Stanford University: SUPREM3,"
URL: http://www-tcad.stanford.edu/tcad/programs/suprem3.html.

[71] C. Ho, J. Plummer, S. Hansen, and R. Dutton, "VLSI Process Modeling—SUPREM III," *IEEE Trans.Electron Devices,* vol. ED-30, no. 11, pp. 1438–1453, 1983.

[72] R. Dutton, "TCAD Tools at Stanford University: SUPREM4,"
URL: http://www-tcad.stanford.edu/tcad/programs/suprem4.html.

[73] Technology Modeling Associates, Inc., Sunnyvale, California, *TMA TSUPREM-4, Two-Dimensional Process Simulation Program, Version 6.5 User's Manual,* 1997.

[74] R. Dutton, "TCAD Tools at Stanford University: PISCES,"
URL: http://www-tcad.stanford.edu/tcad/programs/pisces.html.

[75] S. Beebe, F. Rotella, Z. Sahul, D. Yergeau, G. McKenna, L. So, Z. Yu, K. Wu, E. Kan, J. McVittie, and R. Dutton, "Next Generation Stanford TCAD—PISCES 2ET and SUPREM 007," in *Proc. Intl. Electron Devices Meeting,* pp. 213–216, 1994.

[76] Synopsys (formerly ISE AG), Mountain View, California, *Taurus-TSUPREM4: Industry-Standard Process Simulation Tool.*
URL: http://www.synopsys.com/products/mixedsignal/taurus/process_simu_ds.html.

[77] Synopsys (formerly ISE AG), Mountain View, California, *Taurus-MEDICI: Industry-Standard Device Simulation Tool.*
URL: http://www.synopsys.com/products/mixedsignal/taurus/device_sim_ds.html.

[78] Synopsys, Fremont, USA, *MEDICI User's Manual*, 2003.

[79] Silvaco, Int., Santa Clara, California, *ATHENA: Process Simulation Framework*.
URL: http://www.silvaco.com/products/process_simulation/athena.html.

[80] Silvaco, *ATHENA: 2D Process Simulation Framework*, 1993. User's Manual.

[81] Silvaco, Int., Santa Clara, California, *ATLAS: Device Simulation Framework*.
URL: http://www.silvaco.com/products/device_simulation/atlas.html.

[82] Synopsys (formerly ISE AG), Mountain View, California, *DIOS: Next Generation Process Simulator*.
URL: http://www.synopsys.com/products/acmgr/ise/dios_ds.html.

[83] Synopsys (formerly ISE AG), Mountain View, California, *DESSIS: Next Generation Device Simulator*.
URL: http://www.synopsys.com/products/acmgr/ise/dessis_ds.html.

[84] L. Borucki, H. Hansen, and K. Varahramyan, "FEDSS - A 2D Semiconductor Fabrication Process Simulator," *IBM Journal of Research and Development*, vol. 29, no. 3, pp. 263–276, 1985.

[85] E. Buturla, P. Cottrell, B. Grossman, and K. Salsburg, "Finite-Element Analysis of Semiconductor Devices: The FIELDAY Program," *IBM Journal of Research and Development*, vol. 44, no. 1/2, pp. 142–156, 2000.

[86] M. Pinto, D. Boulin, C. Rafferty, R. Smith, W. C. Jr., I. Kizilyalli, and M. Thoma, "Three-Dimensional Characterization of Bipolar Transistors in a Submicron BiCMOS Technology Using Integrated Process and Device Simulation," in *International Electron Devices Meeting Conference Proceedings*, pp. 923–926, 1992.

[87] D. Adalsteinsson and J. Sethian, "A Level Set Approach to a Unified Model for Etching, Deposition, and Lithography I: Algorithms and Two-Dimensional Simulations," *J.Comp.Phys.*, vol. 120, pp. 128–144, 1995.

[88] D. Adalsteinsson and J. Sethian, "A Level Set Approach to a Unified Model for Etching, Deposition, and Lithography II: Three-Dimensional Simulations," *J.Comp.Phys.*, vol. 122, pp. 348–366, 1995.

[89] M. Law, "FLOODS/FLOOPS Manual,"
URL: http://www.tec.ufl.edu/ flooxs/FLOOXS Manual/Intro.html.

[90] T. Kim, W. Chung, H. Shin, J. Oh, J. Shin, and J. Kong, "Profile Simulation and Stress Analysis for Optimization of an Intermetal Dielectric Deposition (IMD) Process," in *Proc. of the 6th International Conference on VLSI and CAD*, (Piscataway, NJ, USA), pp. 57–60, IEEE, 1999.

[91] A. Yuuki, Y. Matsui, and K. Tachibana, "A Study of Radical Fluxes in Silane Plasma CVD from Trench Coverage Analysis," *Japanese Journal of Applied Physics*, vol. 28, pp. 212–218, Feb 1989.

[92] S. Rauf, "Model for Photoresist Trim Etch in Inductively Coupled CF_4O_2 Plasma," *J.Vac.Sci.Technol.B*, vol. 22, pp. 202–211, Jan 2004.

[93] University Berkeley TCAD Group, Berkeley, California, *LAVA: Lithography Analysis using Virtual Access.*
URL: http://cuervo.eecs.berkeley.edu/Volcano/.

[94] B. Grosman, S. Lachman-Shalem, R. Swissa, and D. Lewin, "Yield Enhancement in Photolithography Through Model-Based Process Control: Average Mode Control," *IEEE Trans.Semiconductor Manufacturing*, vol. 18, pp. 86–93, Feb 2005.

[95] S. Kaplan and L. Karklin, "Calibration of Lithography Simulator by Using Sub-Resolution Patterns," in *Proc. of the SPIE*, vol. 1927, pp. 858–867, International Society for Optical Engineering, 1993.

[96] Z. Zhu and C. Liu, "Anisotropic Crystalline Etching Simulation using a Continuous Cellular Automata Algorithm," in *ASME Symposium on Computer Aided Simulation of MEMS*, Nov 1998.

[97] H. Gummel, "A Self-Consistent Iterative Scheme for One-Dimensional Steady State Transistor Calculations," *IEEE Trans.Electron Devices*, vol. ED-11, pp. 455–465, 1964.

[98] P. Cottrell and E. Buturla, "Steady State Analysis of Field Effect Transistors Via the Finite Element Method," in *International Electron Devices Meeting Conference Proceedings*, pp. 51–54, Dec 1975.

[99] S. Selberherr, A. Schütz, and H. Pötzl, "MINIMOS—A Two-Dimensional MOS Transistor Analyzer," *IEEE Trans.Electron Devices*, vol. ED-27, no. 8, pp. 1540–1550, 1980.

[100] M. Pinto, *PISCES IIB*. Stanford University, 1985.

[101] M. Pinto, "Simulation of ULSI device effects," *Electrochemical Society Proceedings*, vol. 91, no. 11, pp. 43–51, 1991.

[102] S.Wagner, *Small-Signal Device and Circuit Simulation*. Dissertation, Technische Universität Wien, 2005.

[103] *DESSIS Manual*, manual 15, p. 15.535. ISE TCAD Release 10.0, 2005.

[104] W. Hänsch and S. Selberherr, "MINIMOS 3: A MOSFET Simulator that Includes Energy Balance," *IEEE Trans.Electron Devices*, vol. ED-34, no. 5, pp. 1074–1078, 1987.

[105] J. Rollins and J. Choma, "Mixed-Mode PISCES-SPICE Coupled Circuit and Device Solver," *IEEE Trans.Computer-Aided Design*, vol. 7, pp. 862–867, 1988.

[106] S. Selberherr, W. Fichtner, and H. Pötzl, "MINIMOS – A Program Package to Facilitate MOS Device Design and Analysis," in *Numerical Analysis of Semiconductor Devices and Integrated Circuits* (B. Browne and J. Miller, eds.), vol. I, (Dublin), pp. 275–279, Boole Press, 1979.

[107] A. Shibkov and V. Axelrad, "Integrated Simulation Flow for Self-Consistent Manufacturability and Circuit Performance Evaluation," in *2005 International Conference on Simulation of Semiconductor Processes and Devices (SISPAD 2005)*, (Tokyo, Japan), pp. 127–130, 2005.

[108] D. Roulston, "Three Dimensional Effects in Bipolar Transistors using Fast Combined 1d and 2d Numerical Simulation," *Physics-of-Semiconductor-Devices*, vol. 2, pp. 967–973, 1998.

[109] D. Roulston, "CAD of Bipolar Custom Chips using a Coupled Process-Device-Circuit Simulation Package," in *IEEE-1983-Custom-Integrated-Circuits-Confrence*, (New York, NY, USA), pp. 229–232, IEEE, 1983.

[110] R. Strasser, C. Pichler, and S. Selberherr, "VISTA - A Framework for Technology CAD Purposes," in Hahn and Lehmann [205], pp. 450–454.

[111] Institut für Mikroelektronik, Technische Universität Wien, Austria, *VISTA Documentation 1.3-1, VLISP Manual*, 1996.

[112] D. Boning and P. Mozumder, "DOE/Opt: A System for Design of Experiments, Response Surface Modeling, and Optimization Using Process and Device Simulation," *IEEE Trans.Semiconductor Manufacturing*, vol. 7, no. 2, pp. 233–244, 1994.

[113] H. Stippel, E. Leitner, C. Pichler, H. Puchner, E. Strasser, and S. Selberherr, "Process Simulation for the 1990s," *Microelectronics Journal*, vol. 26, no. 2/3, pp. 203–215, 1995.

[114] E. Langer and S. Selberherr, "Three-dimensional Process Simulation for Advanced Silicon Semiconductor Devices," in *Advanced Semiconductor Devices and Microsystems* (T. Lalinský, F. Dubecký, J. Osvald, and Š. Haščík, eds.), (Smolenice, Slovakia), pp. 169–176, 1996.

[115] J. Lorenz, B. Baccus, and W. Henke, "Three-Dimensional Process Simulation," *Microelectronic Engineering*, vol. 34, no. 1, pp. 85–100, 1996.

[116] H. Kosina and S. Selberherr, "A Hybrid Device Simulator that Combines Monte Carlo and Drift-Diffusion Analysis," *IEEE Trans.Computer-Aided Design*, vol. 13, no. 2, pp. 201–210, 1994.

[117] H. Kosina and S. Selberherr, "Technology CAD: Process and Device Simulation," in *Proc. 21st International Conference on Microelectronics*, vol. 2, (Niš, Yugoslavia), IEEE, 1997.

[118] M. Sharma and G. Carey, "Semiconductor Device Simulation Using Adaptive Refinement and Flux Upwinding," *IEEE Trans.Computer-Aided Design*, vol. 8, no. 6, pp. 590–598, 1989.

[119] J. Bürgler, W. Coughran Jr., and W. Fichtner, "An Adaptive Grid Refinement Strategy for the Drift-Diffusion Equations," *IEEE Trans.Computer-Aided Design*, vol. 10, no. 10, pp. 1251–1258, 1991.

[120] V. Senez, T. Hoffmann, and A. Tixier, "Calibration of a Two-Dimensional Numerical Model for the Optimization of LOCOS-Type Isolations by Response Surface Methodology," *IEEE Trans.Semiconductor Manufacturing*, vol. 13, no. 4, pp. 416–426, 2000.

[121] A. Hoessinger, R. Minixhofer, and S. Selberherr, "Full Three-Dimensional Analysis of a Non-Volatile Memory Cell," in *Proceedings International Conference on Simulation of Semiconductor Processes and Devices*, (Munich, Germany), pp. 363–366, 2004.

[122] A. Kolodny, S. Nieh, B. Eitan, and J. Shappir, "Analysis and Modeling of Floating-Gate EEPROM Cells," *IEEE Trans.Electron Devices*, vol. 33, no. 6, pp. 835–844, 1986.

[123] Y. Kurozumi and W. Davis, "Polygonal Approximation by the Minimax Method," *Computer Graphics and Image Processing*, vol. 19, pp. 248–264, 1982.

[124] P. Griffin, S. Crowder, and J. Knight, "Dose Loss in Phosphorus Implants due to Transient Diffusion and Interface Segregation," *Appl.Phys.Lett.*, vol. 67, no. 4, pp. 482–484, 1995.

[125] S. Batra, K. Park, J. Lin, S. Yoganathan, J. Lee, S. Banerjee, S. Sun, and G. Lux, "Effect of Dopant Redistribution, Segregation, and Carrier Trapping in As-Implanted MOS Gates," *IEEE Trans.Electron Devices*, vol. 37, no. 11, pp. 2322–2330, 1990.

[126] J. Sethian and D. Adalsteinsson, "An Overview of Level Set Methods for Etching, Deposition, and Lithography Development," *IEEE Trans.Semiconductor Manufacturing*, vol. 10, no. 1, pp. 167–184, 1997.

[127] P. George and H. Borouchaki, *Delaunay Triangulation and Meshing*. Hermes, 1998.

[128] N. Shigyo, H. Tanimoto, and T. Enda, "Mesh Related Problems in Device Simulation: Treatments of Meshing Noise and Leakage Current," *Solid-State Electron.*, vol. 44, pp. 11–16, 2000.

[129] M. Bächtold, M. Emmenegger, J. Korvink, and H. Baltes, "An Error Indicator and Automatic Adaptive Meshing for Electrostatic Boundary Element Simulations," *IEEE Trans.Computer-Aided Design of Integrated Circuits and Systems*, vol. 16, no. 12, pp. 1439–1446, 1997.

[130] W. Huang and R. Russell, "Moving Mesh Strategy Based on a Gradient Flow Equation for Two-Dimensional Problems," *SIAM J.Sci.Comput.*, vol. 20, no. 3, pp. 998–1015, 1999.

[131] M. Nedjalkov and P. Vitanov, "Iteration Approach for Solving the Boltzmann Equation with the Monte Carlo Method," *Solid-State Electron.*, vol. 32, no. 10, pp. 893–896, 1989.

[132] P. Vitanov and M. Nedjalkov, "Iteration Approach for Solving the Inhomogeneous Boltzmann Equation," in *Numerical Analysis of Semiconductor Devices and Integrated Circuits* (J. Miller, ed.), vol. VII, (Copper Mountain), pp. 55–56, Front Range Press, Boulder, 1991.

[133] P. Antognetti, D. Cavilia, and E. Profumo, "CAD Model for Threshold and Subthreshold Conduction in MOSFET's," *IEEE J.Solid-State Circuits*, vol. SC-17, no. 3, pp. 454–458, 1982.

[134] *INSPECT Manual*, manual 4. ISE TCAD Release 10.0, 2005.

[135] B. Sheu, D. Scharfetter, P. Ko, and M. Jeng, "BSIM: Berkeley Short-Channel IGFET Model for MOS Transistors," *IEEE J.Solid-State Circuits*, vol. SC-22, no. 4, pp. 558–566, 1987.

[136] Y. Cheng, M.-C. Jeng, Z. Liu, J. Huang, M. Chan, K. Chen, P. Ko, and C. Hu, "A Physical and Scalable I-V Model in BSIM3v3 for Analog/Digital Circuit Simulation," *IEEE Trans.Electron Devices*, vol. 44, no. 2, pp. 277–287, 1997.

[137] R. Plasun, C. Pichler, T. Simlinger, and S. Selberherr, "Optimization Tasks in Technology CAD," in Hahn and Lehmann [205], pp. 445–449.

[138] R. Plasun, M. Stockinger, R. Strasser, and S. Selberherr, "Simulation Based Optimization Environment and its Application to Semiconductor Devices," in *Intl. Conf. on Applied Modelling and Simulation*, (Honolulu, Hawaii, USA), pp. 313–316, 1998.

[139] R. Plasun, M. Stockinger, and S. Selberherr, "Integrated Optimization Capabilities in the VISTA Technology CAD Framework," *IEEE Trans.Computer-Aided Design of Integrated Circuits and Systems*, vol. 17, no. 12, pp. 1244–1251, 1998.

[140] G. Gaston and A. Walton, "The Integration of Simulation and Response Surface Methodology for the Optimization of IC Processes," *IEEE Trans.Semiconductor Manufacturing*, vol. 7, no. 1, pp. 22–33, 1994.

[141] C. Pichler, R. Plasun, R. Strasser, and S. Selberherr, "High-Level TCAD Task Representation and Automation," *IEEE J.Technology Computer Aided Design*, May 1997. http://www.ieee.org/journal/tcad/accepted/pichler-may97/.

[142] C. Heitzinger, *Simulation and Inverse Modeling of Semiconductor Manufacturing Processes*. Dissertation, Technische Universität Wien, 2002. URL: http://www.iue.tuwien.ac.at/phd/heitzinger.

[143] C. Heitzinger and S. Selberherr, "An Extensible TCAD Optimization Framework Combining Gradient Based and Genetic Optimizers," in *Proc. SPIE International Symposium on Microelectronics and Assembly: Design, Modeling, and Simulation in Microelectronics*, (Singapore), pp. 279–289, 2000.

[144] R. Strasser, R. Plasun, and S. Selberherr, "Practical Inverse Modeling with SIESTA," in *Simulation of Semiconductor Processes and Devices*, (Kyoto, Japan), pp. 91–94, 1999.

[145] R. Strasser, R. Plasun, M. Stockinger, and S. Selberherr, "Inverse Modeling of Semiconductor Devices," in *Proc. SIAM Conference on Optimization 1999*, p. 77, 1999.

[146] A. Benninghoven, F. Rudenauer, and H. Werner, eds., *Secondary Ion Mass Spectrometry*, (New York), Wiley, 1987.

[147] A. Casel and H. Jorke, "Comparison of Carrier Profiles from Spreading Resistance Analysis and from Model Calculations for Abrubt Doping Structures," *Appl.Phys.Lett.*, vol. 50, no. 15, pp. 989–991, 1987.

[148] T. Clarysse and W. Vandervorst, "A Contact Model for Poisson-Based Spreading Resistance Correction Schemes Incorporating Schottky Barrier and Pressure Effects," *J.Vac.Sci.Technol.B*, vol. 10, no. 1, pp. 413–420, 1992.

[149] Brooks-PRI, Chelmsford, Massachusetts, *Promis MES System*. URL: http://www.brooks-pri.com.

[150] International Business Machines Corp., Armonk, New York, *SiView MES System*. URL: http://houns54.clearlake.ibm.com/solutions/industrial /indpub.nsf/detailcontacts/SiView_Standard.

[151] Applied Materials, Santa Clara, California, *WorkStream MES System*. URL: http://www.appliedmaterials.com/products/workstream.html.

[152] L. Gruber, N. Khalil, D. Bell, and J. Faricelli, *Simulation of Semiconductor Processes and Devices*, ch. Modeling Process and Transistor Variation for Circuit Performance Analysis, pp. 81–84. New York, LLC: Springer-Verlag, 1999.

[153] S. Williams and K. Varahramyan, "A New TCAD-Based Statistical Methodology for the Optimization and Sensitivity Analysis of Semiconductor Technologies," *IEEE Transactions on Semiconductor Manufacturing*, vol. 13, pp. 208–218, May 2000.

[154] V. Nilsen, A. Walton, J. Donnelly, G. Horsburgh, and R. Childs, "Implementation of a TCAD Based System to Aid Process Transfer," in *Proceedings of the 4th International Workshop on Statistical Metrology*, (Piscataway, USA), pp. 54–57, 1999.

[155] A. Stephens, "Avoiding Furnace Slip in the Era of Shallow Trench Isolation," in *Semiconductor Silicon 2002 (9th International Symposium)* (H. Huff, L. Fabry, and S. Kishino, eds.), vol. 2 of *The Electrochemical Society Proceedings Series*, (Pennington, NJ), pp. 774–785, The Electrochemical Society, 2002.

[156] S.-C. Wong, G.-Y. Lee, and D.-J. Ma, "Modeling of Interconnect Capacitance, Delay, and Crosstalk in VLSI," *IEEE Trans.Semiconductor Manufacturing*, vol. 13, no. 1, pp. 108–111, 2000.

[157] M. Bächtold, M. Spasojevic, C. Lage, and P. Ljung, "A System for Full-Chip and Critical Net Parasitic Extraction for ULSI Interconnects using a Fast 3-D Field Solver," *IEEE Trans.Computer-Aided Design of Integrated Circuits and Systems*, vol. 19, no. 3, pp. 325–338, 2000.

[158] Y. Eo, W. Eisenstadt, J. Y. Jeong, and O. Kwon, "A New On-Chip Interconnect Crosstalk Model and Experimental Verification for CMOS VLSI Circuit Design," *IEEE Trans.Electron Devices*, vol. 47, no. 1, pp. 129–140, 2000.

[159] P. Fleischmann, R. Sabelka, A. Stach, R. Strasser, and S. Selberherr, "Grid Generation for Three-Dimensional Process and Device Simulation," in *Simulation of Semiconductor Processes and Devices*, (Tokyo, Japan), pp. 161–166, Business Center for Academic Societies Japan, 1996.

[160] A. Stach, R. Sabelka, and S. Selberherr, "Three-Dimensional Layout-Based Thermal and Capacitive Simulation of Interconnect Structures," in *Proc. 16th IASTED Int. Conf. on Modelling, Identification and Control*, (Innsbruck, Austria), pp. 16–19, 1997.

[161] C. Harlander, R. Sabelka, and S. Selberherr, "A Comparative Study of Two Numerical Techniques for Inductance Calculation in Interconnect Structures," in SISPAD'01 [206], pp. 254–257.

[162] R. Sabelka and S. Selberherr, "SAP — A Program Package for Three-Dimensional Interconnect Simulation," in *Proc. Intl. Interconnect Technology Conference*, (Burlingame, California), pp. 250–252, 1998.

[163] R. Martins, W. Pyka, R. Sabelka, and S. Selberherr, "Modeling Integrated Circuit Interconnections," in *Proc. Intl. Conf. on Microelectronics and Packaging*, (Curitiba, Brazil), pp. 144–151, 1998.

[164] H. Kirchauer and S. Selberherr, "Three-Dimensional Photolithography Simulation," in *Basics and Technology of Electronic Devices* (K. Riedling, ed.), (Grossarl, Austria), pp. 27–31, Gesellschaft für Mikroelektronik, Mar. 1997. Proc. of the Seminar "Grundlagen und Technologie Elektronischer Bauelemente".

[165] H. Kirchauer and S. Selberherr, "Three-Dimensional Photolithography Simulator Including Rigorous Nonplanar Exposure Simulation for Off-Axis Illumination," in *Proc. SPIE Optical Microlithography XI*, vol. 3334, pp. 764–776, 1998.

[166] W. Brown and J. Brewer, eds., *Nonvolatile Semiconductor Memory Technology*. New York: IEEE Press, 1998.

[167] B. Riccò, G. Torelli, M. Lanzoni, A. Manstretta, H. Maes, D. Montanari, and A. Modelli, "Nonvolatile Multilevel Memories for Digital Applications," *Proc.IEEE*, vol. 86, no. 12, pp. 2399–2421, 1998.

[168] S. Lovett, "The Nonvolatile Cell Hidden in Standard CMOS Logic Technologies," *IEEE Trans.Electron Devices*, vol. 48, no. 5, pp. 1017–1018, 2001.

[169] P. Pavan, R. Bez, P. Olivo, and E. Zanoni, "Flash Memory Cells – An Overview," *Proc.IEEE*, vol. 85, no. 8, pp. 1248–1271, 1997.

[170] W. Hänsch and C. Schmeiser, "Hot Electron Transport in Semiconductors," *Zeitschrift für angewandte Mathematik und Physik*, vol. 40, pp. 440–455, 1989.

[171] M. Lenzlinger and E. Snow, "Fowler-Nordheim Tunneling into Thermally Grown SiO_2," *J.Appl.Phys.*, vol. 40, no. 1, pp. 278–283, 1969.

[172] R. Duane, A. Concannon, P. O'Sullivan, M. O'Shea, and A. Mathewson, "Extraction of Coupling Ratios for Fowler-Nordheim Programming Conditions," *Solid-State Electron.*, vol. 45, no. 2-3, pp. 235–242, 2001.

[173] J. M. Caywood, C. J. Huang, and Y. J. Chang, "A Novel Nonvolatile Memory Cell Suitable for Both Flash and Byte-Writable Applications," *IEEE Trans.Electron Devices*, vol. 49, no. 5, pp. 802–807, 2002.

[174] C. Huang, "A Novel P-Channel Flash Electrically-Erasable Programmable Read-Only Memory(EEPROM) Cell with Oxide-Nitride-Oxide (ONO) as Split Gate Channel Dielectric," *Jpn. J. Appl. Phys.*, vol. 40, no. 4B, pp. 2943–2947, 2001.

[175] H. Yu, Y. Hou, M. Li, and D. Kwong, "Hole Tunneling Current Through Oxynitride/Oxide Stack and the Stack Optimization for p–MOSFETs," *IEEE Electron Device Lett.*, vol. 23, no. 5, pp. 285–287, 2002.

[176] R. Fowler and L. Nordheim, "Electron Emission in Intense Electric Fields," *Proc.Roy.Soc.A*, vol. 119, pp. 173–181, 1928.

[177] M. Lenzinger and E. Snow, "Fowler-Nordheim Tunneling into Thermally Grown SiO," *J. Appl. Phys.*, vol. 40, no. 1, pp. 278–283, 1969.

[178] A. Gehring, *Simulation of Tunneling in Semiconductor Devices*. Dissertation, Technische Universität Wien, 2003.
URL: http://www.iue.tuwien.ac.at/phd/gehring.

[179] A. Hössinger, T. Binder, W. Pyka, and S. Selberherr, "Advanced Hybrid Cellular Based Approach for Three-Dimensional Etching and Deposition Simulation," in SISPAD'01 [206], pp. 424–427.

[180] A. Hössinger, J. Cervenka, and S. Selberherr, "A Multistage Smoothing Algorithm for Coupling Cellular and Polygonal Datastructures," in *Proc. 2003 Intl. Conf. on Simulation of Semiconductor Processes and Devices*, pp. 259–262, 2003.

[181] H. Kirchauer and S. Selberherr, "Three-Dimensional Photolithography Simulator Including Rigorous Nonplanar Exposure Simulation for Off-Axis Illumination," in *Optical Microlithography XI* (L. V. der Hove, ed.), pp. 764–776, SPIE, 1998.

[182] H. Kirchauer and S. Selberherr, "Rigorous Three-Dimensional Photolitography Simulation Over Nonplanar Structures," in *26th European Solid State Device Research Conference* (G. Baccarani and M. Rudan, eds.), (Gif-sur-Yvette Cedex, France), pp. 347–350, Editions Frontieres, 1996.

[183] W. Pyka, H. Kirchauer, and S. Selberherr, "Three-Dimensional Resist Development Simulation – Benchmarks and Integration with Lithography," *Microelectronic Engineering*, vol. 53, no. (1-4), pp. 449–452, 2000.

[184] A. Schenk, "Rigorous Theory and Simplified Model of the Band-to-Band Tunneling in Silicon," *Solid-State Electron.*, vol. 36, no. 1, pp. 19–34, 1993.

[185] J. Doyle, "A Thick Polysilicon Three-State Fuse," *Motorola Technical Developments*, vol. 3, pp. 31–32, 1983.

[186] O. Kim, "CMOS Trimming Circuit Based on Polysilicon Fusing," *Electronic Letters*, vol. 34, pp. 355–356, Feb 1998.

[187] D. Nickel, "Element Trimming Fusible Link," *IBM Technical Disclosure Bulletin*, vol. 26, no. 8, p. 4415, 1984.

[188] J. Lloyd and M. Polcari, "Polysilicon Fuse," *IBM Technical Disclosure Bulletin*, vol. 24, no. 7A, p. 3442, 1981.

[189] D. Greve, "Programming Mechanism of Polysilicon Resistor Fuses," *IEEE Trans. Electron Devices*, vol. 29, no. 4, pp. 719–724, 1982.

[190] Y. Fukuda, S. Kohda, K. Masuda, and Y. Kitano, "A New Fusible-Type Programmable Element Composed of Aluminum and Polysilicon," *IEEE Trans. Electron Devices*, vol. 33, no. 2, pp. 250–253, 1986.

[191] A. Kalnitsky, I. Saadat, A. Bergemont, and P. Francis, "CoSi$_2$ Integrated Fuses on Poly Silicon for Low Voltage 0.18μm CMOS Applications," in *International Electron Devices Meeting Conference Proceedings*, pp. 765–768, 1999.

[192] D. Greve, "Programming Mechanism of Polysilicon Fuse Links," in *International Electron Devices Meeting Conference Proceedings*, pp. 70–73, 1981.

[193] S. Das and S. Lahiri, "Transient Response of Polysilicon Fuse-Links for Programmable Memories and Circuits," in *Semiconductor Devices (Proc. SPIE Vol.2733)*, (New Delhi, India), pp. 232–234, 1995.

[194] A. McConnel, S. Uma, and K. Goodson, "Thermal Conductivity of Doped Polysilicon Layers," in *Proc. of the Int. Conference on Heat Transfer and Transport Phenomena in Microscales* (G. Celata and et. al., eds.), (Begell House, New York), pp. 413–419, 2000.

[195] R. Sabelka, "A Finite Element Simulator for Three-Dimensional Analysis of Interconnect Structures," *Microelectronics Journal*, vol. 32, no. 2, pp. 163–171, 2001.

[196] M. Winter, "WebElements,"
URL: http://www.webelements.com/webelements/index.html.

[197] National Institute of Standards and Technology, Gaithersburgh, Maryland, *NIST Webbook*.
URL: http://webbook.nist.gov/chemistry/name-ser.htm.

[198] Automation Creations Inc., Blacksburg, Virginia, *MatWeb: Material Property Data*.
URL: http://www.matweb.com.

[199] C. Vahlas, P.-Y. Chevalier, and E. Blanquet, "A Thermodynamic Evaluation of Four Si-M (M = Mo, Ta, Ti, W) Binary Systems," *CALPHAD: Comput. Coupling Phase Diagrams Thermochem.*, vol. 13, no. 3, pp. 273–292, 1989.

[200] M. Mandurah and K. Saraswat, "A Model for Conduction in Polycrystalline Silicon-Part I: Theory," *IEEE Trans. Electron Devices*, vol. 28, no. 10, pp. 1163–1171, 1981.

[201] C. Jung, *Über die Beziehung der Psychotherapie zur Seelsorge*, p. 362. 1932.

[202] T. Feudel, W.Fichtner, N.Strecker, R.Zingg, G.Dallmann, and E.Döring, "Improved Technology Understanding through Using Process Simulation and Measurements," in *Proceedings of the Simulation of Semiconductor Processes and Devices - SISDEP 93*, vol. 5, (Springer, Wien, Austria), pp. 217–220, 1993.

[203] Government Electronics & Information Technology Association, Arlington, Virginia, *Electronic Design Interchange Format (EDIF) Documentation*.
URL: http://www.eigroup.org.

[204] A. Goda and K. Hane, "A Method of 1-Dimensional Device Simulation by Boltzmann Transport Equation," in *NASECODE VI - Numerical Analysis of Semiconductor Devices and Integrated Circuits* (J. Miller, ed.), (Dublin), pp. 360–365, Boole Press, 1989.

[205] W. Hahn and A. Lehmann, eds., *Proc. 9th European Simulation Symposium*, (Passau, Germany), Society for Computer Simulation International, 1997.

[206] *Proc. Simulation of Semiconductor Processes and Devices*, (Athens, Greece), 2001.

Own Publications

[1] G. Roehrer, M. Knaipp, R. Minixhofer, and H. Noll, "Investigations of Sidewall Effects in Bipolar Transistors," in *Proc. Microelectronics, Devices and Materials, MIDEM*, pp. 133–138, 1999.

[2] C. Harlander, R. Sabelka, R. Minixhofer, and S. Selberherr, "Three-Dimensional Transient Electro-Thermal Simulation," in *5th THERMINIC Workshop*, (Rome, Italy), pp. 169–172, Oct. 1999.

[3] R. Minixhofer, "Why Bother about Substrate Currents? Risks of Substrate Currents in Smart Power Chips," in *Workshop on Substrate Current Effects in Smart-Power and Mixed-Signal ASICs*, (Cork, Ireland), pp. 5–10, Sept. 1999. (invited).

[4] R. Minixhofer, G. Roehrer, and S. Selberherr, "Implementation of an Automated Interface for Integration of TCAD with Semiconductor Fabrication," in *Proceedings Simulation in Industry-14th European Simulation Symposium and Exhibition*, (Dresden, Germany), pp. 70–74, Oct. 2002.

[5] R. Minixhofer, S. Holzer, C. Heitzinger, J. Fellner, T. Grasser, and S. Selberherr, "Optimization of Electrothermal Material Parameters using Inverse Modeling," in *Proceedings 33rd European Solid-State Devices Research Conference*, (Estoril, Portugal), pp. 363–366, Sept. 2003.

[6] J. Cervenka, A. Hoessinger, R. Minixhofer, T. Grasser, and S. Selberherr, "Dreidimensionale Modellierung Elektronischer Bauelemente," in *Beiträge der Informationstagung Mikroelektronik*, (Vienna, Austria), pp. 377–382, Oct. 2003.

[7] S. Holzer, R. Minixhofer, C. Heitzinger, J. Fellner, T. Grasser, and S. Selberherr, "Extraction of Material Parameters Based on Inverse Modeling of Three-Dimensional Interconnect Structures," in *Proceedings 9th THERMINIC Workshop*, (Aix-en-Provence, France), pp. 263–268, Sept. 2003.

[8] G. Leonardelli, G. Roehrer, and R. Minixhofer, "A Novel System for Fully Automated Creation of Layout Documentation and Test Programs for Electrical Test Structures," in *Proceedings IEEE/SEMI Advanced Semiconductor Manufacturing Conference and Workshop*, (Boston, USA), pp. 205–207, May 2004.

[9] S. Holzer, R. Minixhofer, C. Heitzinger, J. Fellner, T. Grasser, and S. Selberherr, "Extraction of Material Parameters Based on Inverse Modeling of Three-Dimensional Interconnect Fusing Structures," *Microelectronics Journal*, vol. 35, no. 2/3, pp. 805–810, 2004.

[10] M. Knaipp, G. Roehrer, R. Minixhofer, and E. Seebacher, "Investigations on the High Current Behavior of Lateral Diffused High-Voltage Transistors," *IEEE Trans.Electron Devices*, vol. 51, pp. 1711–1720, Oct. 2004.

[11] A. Hoessinger, R. Minixhofer, and S. Selberherr, "Full Three-Dimensional Analysis of a Non-Volatile Memory Cell," in *Proceedings International Conference on Simulation of Semiconductor Processes and Devices*, (Munich, Germany), pp. 363–366, 2004.

[12] R. Minixhofer and D. Rathei, "Using TCAD for Fast Analysis of Misprocessed Wafers and Yield Excursions," in *Proceedings IEEE/SEMI Advanced Semiconductor Manufacturing Conference and Workshop*, (Munich, Germany), pp. 1–3, Apr. 2005.

[13] R. Minixhofer, "Semiconductor Process Simulation," in *Forum Gesellschaft für Mikroelektronik*, (Vienna, Austria), p. 31, Mar. 2005. (invited).

www.ingramcontent.com/pod-product-compliance
Lightning Source LLC
LaVergne TN
LVHW022312060326
832902LV00020B/3417